PSYCHOLOGICAL BASIS
OF SENSORY EVALUATION

PSYCHOLOGICAL BASIS OF SENSORY EVALUATION

Edited by

R. L. McBRIDE

*Sensometrics Pty Ltd, 357 Military Road,
Mosman, NSW 2088, Australia*

and

H. J. H. MacFIE

*AFRC Institute of Food Research, Reading Laboratory,
Shinfield, Reading, UK*

ELSEVIER APPLIED SCIENCE
LONDON and NEW YORK

ELSEVIER SCIENCE PUBLISHERS LTD
Crown House, Linton Road, Barking, Essex IG11 8JU, England

Sole Distributor in the USA and Canada
ELSEVIER SCIENCE PUBLISHING CO., INC.
655 Avenue of the Americas, New York, NY 10010, USA

WITH 4 TABLES and 24 ILLUSTRATIONS

© 1990 ELSEVIER SCIENCE PUBLISHERS LTD

British Library Cataloguing in Publication Data

Psychological basis of sensory evaluation.
1. Man. Gustatory perception. Psychological aspects
I. McBride, R. L. MacFie, J. H. J.
612'.87'019

ISBN 1-85166-453-X

Library of Congress Cataloging in Publication Data

Psychological basis of sensory evaluation/edited by R. L. McBride and
H. J. H. Macfie.
  p.   cm.
Includes bibliographical references.

ISBN 1-85166-453-X (U.S.)
1. Food — Sensory evaluation — Psychological aspects.
2. Taste — Psychological aspects.   I. McBride, R. L.   II. Macfie, H. J. H.
TX546.P78 1990
664'.07 — dc20

No responsibility is assumed by the Publisher for any injury and/or damage to persons or property as a matter of products liability, negligence or otherwise, or from any use or operation of any methods, products, instructions or ideas contained in the material herein.

Special regulations for readers in the USA

This publication has been registered with the Copyright Clearance Center Inc. (CCC), Salem, Massachusetts. Information can be obtained from the CCC about conditions under which photocopies of parts of this publication may be made in the USA. All other copyright questions, including photocopying outside the USA, should be referred to the publisher.

All rights reserved. No part of this publication may be reproduced, stored in a retrieval system, or transmitted in any form or by any means, electronic, mechanical, photocopying, recording, or otherwise, without the prior written permission of the publisher.

Phototypesetting by Tech-Set, Gateshead, Tyne & Wear.
Printed in Great Britain at the University Press, Cambridge.

Preface

The origins of this book go back to an SCI symposium with the same title organised by one of the authors (HM) at which the senior author was the principal speaker. Although the meeting was very well attended and generated a great deal of interest, it was clear that the topic had not been covered and we determined to produce a text on the subject.

The need for such a book is easy to justify. Throughout the global industrial community there is a continuing need for the human assessment of food, drink, perfumery, pharmaceutical and domestic products. These assessments are being organised by sensory personnel with very little formal training in the topic for the simple reason that very few academic institutions provide any. Industrial courses do fill the gap but are usually constrained by time limits to concentrate on 'how to do it' rather than 'why do it this way' or 'what can go wrong'.

The continual pressure from clients for results also serves to give the statistical aspects of sensory evaluation a high priority, and there are now a number of sophisticated multivariate techniques that seemingly cope with all forms of errant behaviour by panellists; however we must not forget that these measuring instruments — people — are easily influenced by careless experimental procedures.

The aim of this book is therefore to summarise, in a way that is comprehensible to all practitioners of sensory evaluation, the substantial body of research that has been carried out by psychologists into aspects of sensory perception.

If we are to use a particular aspect of human perception in our testing we should understand how it works and develops throughout the life cycle. The first three chapters address the development of taste perception, the effects of ageing on sensory functioning and the psychophysical aspects of perception of complex taste stimuli.

Similarly if the psychological basis of a sensory test is unsound no amount of fancy statistical analysis will put it right. The next three chapters, which form the heart of the text, address the psychological issues affecting sensory procedures and propose technical solutions.

The use of sensory evaluation in product development and as an input into a model of food choice is being questioned by psychologists. The final three chapters record some of this thinking and some new approaches to these topics.

We hope this text will serve the double purpose of encouraging sensory workers to follow the progress of research in psychobiology and at the same time tempt more psychologists to enter the sensory community and tackle the many outstanding problems and psychological issues that remain unresolved.

Finally, it is a pleasure to acknowledge the assistance of many colleagues who have helped us in designing the text, most particularly the authors of the chapters who have worked so hard in producing and revising their submissions; and of course the publishers for their encouragement and help.

H. J. H. MACFIE

Contents

Preface v

List of Contributors ix

1. Early Development of Taste Perception 1
   B. J. COWART and G. K. BEAUCHAMP

2. Effects of Aging on Sensory Functioning: Implications for Dietary Selection 19
   C. MURPHY and M. M. GILMORE

3. The Perception of Complex Taste Stimuli 41
   J. H. A. KROEZE

4. Applications of Experimental Psychology in Sensory Evaluation 69
   H. LAWLESS

5. Integration Psychophysics 93
   R. L. MCBRIDE and N. H. ANDERSON

6. Cognitive Aspects of Difference Testing and Descriptive Analysis: Criterion Variation and Concept Formation . . 117
   M. O'MAHONY

7. Attitudes and Beliefs as Determinants of Food Choice . . 141
   R. SHEPHERD

8. Designing Products for Individual Customers . . . 163
   D. A. BOOTH

9. Three Generations of Sensory Evaluation 195
   R. L. MCBRIDE

Index 207

List of Contributors

N. H. ANDERSON
Department of Psychology, University of California, San Diego, La Jolla, California 92093, USA

G. K. BEAUCHAMP
Monell Chemical Senses Center, 3500 Market Street, Philadelphia, Pennsylvania 19104, USA

D. A. BOOTH
School of Psychology, University of Birmingham, Edgbaston, Birmingham B15 2TT, UK

B. J. COWART
Monell Chemical Senses Center, 3500 Market Street, Philadelphia, Pennsylvania 19104, USA

M. M. GILMORE
Department of Psychology, Brown University, Providence, Rhode Island 02912, USA

J. H. A. KROEZE
Laboratory of Psychology, University of Utrecht, PO Box 80.140, 3508TC, Utrecht, The Netherlands

H. LAWLESS
Department of Food Science, New York State College of Agriculture and Life Sciences, Cornell University, Ithaca, NY 14853, USA

M. O'MAHONY
Department of Food Science and Technology, University of California, Davis, California 95616, USA

R. L. MCBRIDE
Sensometrics Pty Ltd, 357 Military Road, Mosman, NSW 2088, Australia

C. MURPHY
Department of Psychology, San Diego State University, San Diego, California 92182-0551, USA

R. SHEPHERD
AFRC Institute of Food Research, Reading Laboratory, Shinfield, Reading RG2 9AT, UK

1
Early Development of Taste Perception

BEVERLY J. COWART & GARY K. BEAUCHAMP
*Monell Chemical Senses Center, 3500 Market Street, Philadelphia,
Pennsylvania 19104, USA*

INTRODUCTION

This chapter provides an overview of what is known about the development of human taste perception by focusing on the methods that have been used to assess sensitivities and preferences in infants and young children. Particular attention is given to the interpretation of response measures obtained in studies of early chemosensory development. In this context, we present examples from the literature as well as from our own recent work that serve to illustrate how the conclusions drawn about infants' and children's perceptions can vary with the methods employed to ask questions about those perceptions.

NEWBORNS

Selected review of the literature
Published studies of taste perception in early infancy first began to appear in the late 1800s, and this area attracted substantial research attention in the 1970s (Cowart, 1981). As a result, there is a larger body of studies relating to this aspect of the development of flavor perception than to any other. Table 1 comprises a list of the principal response measures that have been used to assess neonatal taste perception. With one exception, all of these responses are believed to reflect hedonic reactions. That is, it has been assumed they provide some measure of preference or acceptance or pleasure. When infants exhibit affective

TABLE 1
Measures used to assess taste perception in newborns

Tongue movements
Autonomic reactivity (e.g. heart or respiratory rate)
Facial expressions
Differential ingestion (i.e. measures of intake)
Sucking patterns

responses to taste stimuli, it can obviously be inferred that they are sensitive to them; failure to observe such responses does not, however, indicate whether the infant failed to detect the stimulus or was simply hedonically indifferent to it. Consequently, the ability to make statements about the absolute taste sensitivities of newborn infants is still constrained.

The single, generally acknowledged exception to hedonically motivated responses among those listed is one form of tongue movement, the lateral tongue reflex, which was first described by Weiffenbach and Thach (1973). This reflex is elicited by the application of microdrops of fluid to the newborn's tongue, and neither its form nor frequency is affected by qualitative changes in the fluid. Thus, the occurrence of the response appears to be solely dependent on sensory mechanisms. Using an adaptation paradigm, Weiffenbach and Thach (1973) and Weiffenbach (1977) have demonstrated differential responding to water, glucose and sodium chloride (NaCl) in newborns. Specifically, following adaptation of the response to drops of water, responding to glucose (0·28M, 0·55M and 2·77M) and to saline (2M) remains intact. In addition, adaptation to 2M glucose does not cross-adapt the response to 2M saline, suggesting that different receptor mechanisms underlie the newborn's responses to these tastants; however, 2M salt is undoubtedly a trigeminal as well as a gustatory stimulus, producing pain/irritant sensations (Abrahams *et al.*, 1937; Kawamura *et al.*, 1968; Green & Gelhard, 1989). As a consequence, distinct functional taste receptors for sweet and salt are not necessarily implied by this finding. Unfortunately, differential responding has only been reliably observed with concentrated taste solutions, and the utility of this response as a measure of absolute gustatory sensitivity is questionable.

There is one other form of response listed in Table 1 for which the characterization of 'hedonic' might be questioned, and that is autonomic reactivity. Crook and Lipsitt (1976) and Lipsitt (1977) have published the only modern studies of infant taste perception using such measures, and their research was limited to sweet taste responsivity. In both studies, heart rate was found to be positively correlated with the concentration of sucrose presented to infants. Lipsitt (1977) proposed that the heart rate response is a signal of 'the joy that the infant . . . derives from [the] 'pleasure of sensation' (p. 138). Unpublished data obtained by P. A. Daniel (see Cowart, 1981) apparently suggest that salty solutions can have the reverse effect on heart rate and, thus, may represent an example of a negative form of this response. In the absence of more substantive evidence, however, the possibility cannot be discounted that the heart rate response reflects arousal rather than pleasure and could provide a measure of gustatory sensitivity, as opposed to preference, in infancy.

Consistent with that interpretation, heart rate and other autonomic responses (primarily respiratory rate) have been used productively to study the early development of olfactory sensitivity. The direction of these responses does not vary as a function of the quality of odor stimuli (Engen *et al.*, 1963), but response magnitude appears to be sensitive to such stimulus parameters as concentration and repeated presentation. As a result, measures of autonomic reactivity have allowed for a more sophisticated characterization of olfactory function in newborns than is available for gustatory function, including the assessment of olfactory thresholds (Lipsitt *et al.*, 1963), habituation (Engen & Lipsitt, 1965) and cross-adaptation (Rovee, 1972).

The remaining measures listed in Table 1 have been used more extensively in infant taste studies. Although care should be taken in assigning hedonic meaning to specific parameters of these measures, facial expressions, differential ingestion (intake) and sucking patterns all seem to largely reflect hedonic reactions and have yielded generally consistent results. There are, however, some apparent inconsistencies that have yet to be fully resolved.

Facial expressions, suggestive of contentment and liking or discomfort and rejection, were used to assess the newborn's responsiveness to taste stimuli in some of the earliest investigations of human taste development (e.g. Preyer, 1882/1901; Shinn, 1900; Peterson & Rainey, 1910). The most extensive studies of these so-called gusto-facial responses have been conducted by Steiner (1973, 1977, 1979), who has reported that consistent, quality-specific facial expressions can be elicited in the first few hours of life by sweet (sucrose, 0·73M), sour (citric acid, 0·12M) and bitter (quinine sulfate, 0·003M) stimuli. Because the stimuli used to elicit these responses were quite concentrated and might, in some cases, have elicited a trigeminal as well as gustatory response, differences among the responses to each taste could be at least partially reflective of differences in the extent of trigeminal involvement. In only one study have Steiner and his colleagues (Ganchrow et al., 1983) examined responses to less concentrated taste stimuli or provided data on the frequency of occurrence of specific facial features. Stimuli in that study were distilled water, sucrose (0·1 and 1·0M), quinine hydrochloride (0·0001M) and an additional bitter tastant, urea (0·15 and 0·25M). The quinine, 1·0M sucrose and 0·25M urea elicited more responses than the other stimuli, and the patterns of response to the quinine and more concentrated urea differed from that elicited by 1·0M sucrose. On the other hand, there was considerable overlap in the frequency distributions of facial features elicited by all of the stimuli, and responses elicited by the weaker concentrations of sucrose and urea were not substantially different from those observed following water presentations.

A more detailed description of newborns' gusto-facial expressions has recently been provided by Rosenstein and Oster (1988). These authors used a modification for infants' faces of Ekman and Friesen's (1978) Facial Action Coding System, an anatomically based system that specifies facial movements in terms of minimally distinguishable actions of the facial muscles. The sweet, sour and bitter stimuli they presented were the same as those used by Steiner (1973, 1977, 1979), and they also included a salt stimulus (0·73M NaCl). Consistent with Steiner's reports, analyses of the patterns of full-face expressions indicated that the newborns discriminated sweet from non-sweet solutions and, among non-sweet solutions, between the sour and bitter taste stimuli. Rosenstein and Oster's analyses also demonstrated, however, that taste-elicited facial expressions are not as highly stereotyped as suggested in Steiner's reports; that is, there appears to be considerable overlap in the distribution of not only discrete facial actions but also full-face configurations elicited by the different taste stimuli, as well as considerable individual variation in facial responses. Moreover, although the authors found that hedonically negative facial components were more common in response to all three non-sweet tastes than in response to sweet, untrained observers found only the responses to the bitter tastant to be highly negative.

Even with their fine-grained analysis and a concentrated salt stimulus (0·73M), Rosenstein and Oster (1988) were not able to identify a distinctive facial response to salt. As noted, affectively negative facial actions (specifically, in the brow and mid-facial regions) did occur somewhat more often in response to salt than to sugar. These actions were also present in the initial seconds of the infants' responses to sweet, however, and the authors did not assess responses to the water diluent itself. Thus, the bases and meaning of the limited responses to the salt stimulus are unclear.

Differential ingestion would seem to be the most self-evident measure of taste preference in infancy, and it has been used more frequently than any other response. There are, however, several limitations associated with this measure. First, because multiple stimuli cannot be presented simultaneously to human infants, thereby providing a true choice situation, the measure should probably be viewed as one of relative acceptance rather than of preference. More critically, the number and range of stimulus concentrations that it is either feasible or ethical to present to infants in an unrestricted feeding situation are limited. Finally, the strength of the sucking reflex in young infants is quite strong, which may mitigate against clear rejection of a taste stimulus. In fact, although intake studies have repeatedly demonstrated strong acceptance of sweet tastes by infants at birth and shown that newborns respond to even dilute sweet tastes and can differentiate varying degrees of sweetness (e.g. Desor *et al.*, 1973, 1977; Maller & Desor, 1973), they have generally failed to demonstrate differential responding to sour or bitter tastes as well as to salt. In the single exception to this generalization, Desor *et al.* (1975) used responses to a mild sweet solution rather than plain water as the baseline measure (ensuring the infants were capable of responding at levels both higher and lower than baseline) and presented salty (NaCl), sour (citric acid) and bitter (urea) tastes in this sweet base. Under these circumstances, relative suppression of intake was observed with the addition of citric acid concentrations at and above 0·003M, but no significant changes in intake occurred in response to the addition of either NaCl (0·05 – 0·20M) or urea (0·18 – 0·48M). Absence of differential responding to salt is consistent with results from the studies of facial expression described above, but the apparent indifference to bitter is not. Three points should be noted, however. First, for adults urea is a much less intensive bitter stimulus than is quinine and seems to fall into a separate class of bitter stimuli that may be coded by a different receptor mechanism (McBurney *et al.*, 1972). Second, although Ganchrow *et al.* (1983) found that 0·25M urea elicited facial actions that included apparently negative components similar to those elicited by a quinine stimulus, the urea also elicited more apparently positive facial actions, such as sucking and licking, than did the quinine. Third, when taste stimuli are mixed, the intensity of each of the component tastes tends to be suppressed (interestingly, sour taste intensity may be less susceptible to suppression than the other primary tastes; see Bartoshuk, 1975). Thus, the discrepancy between findings concerning bitter responsivity in newborns could partially reflect differences in the type and/or intensity of bitter stimuli presented.

Finally, the study of sucking patterns has been utilized as an alternative to the measurement of intake *per se*. Techniques for recording multiple parameters of sucking in response to small stimulus volumes can potentially allow for the study of responses to sapid substances that one would be reluctant to have infants consume in any quantity. On the other hand, the relative significance of individual parameters of the sucking response and the relationships among different parameters are not well understood. Presentation of minute amounts of sweet solution has been shown to lead to concentration-dependent increases in the number and length of bursts of sucking activity and to decreases in the time between bursts, all of which are intuitively consistent with a positive hedonic reaction (Crook, 1977, 1978). Seemingly paradoxically, however, the rate of sucking within bursts also decreases (Crook & Lipsitt, 1976; Crook, 1977). Of these parameters, only burst length has been examined following presentation of salt solutions, and even a weak salt solution (0·1M) has been reported to significantly decrease the length of sucking bursts (Crook, 1978). Thus, in contrast to facial expression and intake measures, this response measure seems to suggest that newborns can detect the presence of salt and that they find it hedonically negative. Responses to sour and bitter stimuli have not been evaluated using this kind of procedure.

Summary

Although each of the measures used in studies of newborn taste responses has its limitations, the convergence of results from different measures of sweet taste perception gives confidence in the conclusion that infants are quite sensitive to this taste and display an innate preference for it. Similarly, even though it has been studied less extensively, measures of intake and facial expression suggest that sour taste perception may be reasonably well developed at birth and that newborns reject this taste. The highly concentrated sour stimuli that have been used to elicit facial expressions do, however, pose some interpretive problems, and additional studies of sour taste responsivity are necessary before strong conclusions can be drawn with regard to sour perception in early development.

Neonatal responses to bitter are more problematic. Newborns respond with highly negative facial expressions to concentrated quinine, but not necessarily to urea. Whether this reflects relative insensitivity and/or indifference to low levels of bitterness, or to particular bitter stimuli, is unclear. Further studies using a variety of bitter stimuli and additional measures, such as sucking patterns and autonomic reactions, could resolve this question.

The newborn's response, or lack thereof, to salt is perhaps most puzzling. Given the ambiguous nature of most of the responses that have been observed, and the evidence for postnatal development of salt taste reception in other mammalian species (Hill & Almli, 1980; Ferrel et al., 1981; Hill et al., 1982; Mistretta & Bradley, 1983), it seems likely that humans are relatively insensitive to salt at birth. Although some measures suggest that newborns react negatively to salt stimulation, it is premature to conclude that saltiness is an inherently negative taste quality. More extensive exploration of the various parameters of sucking responses to a range of salt concentrations may shed further light on this issue.

OLDER INFANTS AND YOUNG CHILDREN

Selected review of the literature

There are relatively few studies of taste perception in older infants and children as compared to neonates, and a large proportion of those that have been reported were based on small samples and not described in detail. For all practical purposes, the measures obtained have been limited to intake (i.e. differential ingestion) and paired-comparison or rank-ordering of taste stimuli to determine preference, and with few exceptions, only responses to sweet and salty tastes have been assessed.

Infants beyond the neonatal period (1–24 months) have been most neglected in studies of taste development. In part this is probably due to the lack of clearly appropriate measures of either preference or sensitivity for subjects in this age range. Differential ingestion has been the default option in almost all studies. There are, however, certain problems encountered with older infants that do not arise with neonates. For example, they may be less willing to accept unfamiliar bottles or food from an unfamiliar person, and as Filer (1978) has noted, the amount of any particular food eaten in a natural feeding situation may depend as much on the mother's 'mechnical skill... and determination to feed her infant' (p. 7) as on the infant's preference. Finally, because intake has been the only measure obtained from older infants, there is no cross-validation of the meaningfulness of this measure. Nonetheless, a few notable findings, suggestive of developmental/experiential changes in taste preferences during infancy, have been reported.

First, in the case of salt taste, intake measures obtained in two studies (see Beauchamp et al., 1986) suggest that preferential ingestion of salt water relative to plain water emerges at approximately four months of age. Beauchamp et al. (1986) and Cowart and Beauchamp (1986a) have argued that experience with salty tastes probably does not play a major role in the shift from apparent indifference to salt at birth to acceptance in later infancy; rather, this change in response may reflect postnatal maturation of central and/or peripheral mechanisms underlying salt taste perception, allowing for the expression of a largely unlearned preference for saltiness. On the other hand, there is some evidence that, at least by 6 months of age, frequency of dietary exposure to high sodium foods (though not total dietary sodium) may affect the degree of preference for salted versus unsalted cereal. In a study of ten, 6-month-old infants, Harris and Booth (1987) determined the difference between the amount of salted cereal (100 mg NaCl per 100 g prepared weight) eaten during feedings on two days and the amount of unsalted cereal eaten on two different days and used that as a measure of salt preference. They found relative preference to be significantly correlated with the number of times the infants were exposed to foods containing at least the amount of sodium in the salted cereal during the week preceding testing; none of the infants actually rejected the salted cereal relative to the unsalted one, however, and all but one exhibited some preference for it by consuming more salted than unsalted cereal.

Similarly, Beauchamp and Moran (1982) found evidence for effects of dietary experience on the expression of sweet preferences in 6-month-olds. They assessed intake of sweetened versus unsweetened water in the same infants at birth and at 6 months. Although the infants preferred sweetened water at both ages, only the 6-month-olds who had received sugared water as part of their diet continued to exhibit the same degree of preference for sweet in water that they had at birth.

Dietary experience — in particular, familiarity with specific foods and with tastes in specific food contexts — has also been shown to play an important role in the preferences of preschool children (Birch, 1979a; Birch & Marlin, 1982; Beauchamp & Moran, 1985; Cowart & Beauchamp, 1986b). The other major finding concerning taste preferences in early childhood has been that young children tend to exhibit preferences for more concentrated sugar and salt than do adults, although the bases for these extreme preferences have yet to be elucidated (see Cowart & Beauchamp, 1986a; Beauchamp & Cowart, 1987). In virtually all studies of this age group, a paired-comparison or rank-ordering procedure has been employed. Both of these procedures have been found to yield results consistent with those obtained via relative intake measures (Birch, 1979b; Cowart & Beauchamp, 1986b), providing some confidence in their meaningfulness as measures of children's preferences. There are, however, problems that may arise when these techniques are used with young children. The remainder of this chapter focuses on those we have encountered in our studies of salt preferences in early childhood, their potential impact on the results that are obtained, and the modified, paired-comparison procedure we have developed in an attempt to circumvent some of these difficulties.

A preference technique for use with preschoolers

In designing assessment techniques to be used with young children, it is important that the communicative, cognitive (attention and memory) and behavioral capacities of the child be taken into account. Failure to understand a task, to attend to it and/or to remember sequentially presented stimuli about which relative judgments are to be made may lead to arbitrary, although possibly systematic, patterns of response. The fact that young children often exhibit marked response biases has been demonstrated by Engen and Katz (cited in Engen, 1974) in a study of olfactory preferences. They found that at 4 years of age, significantly more children indicated they liked the smell of butyric acid when asked if it smelled 'pretty' than when asked if it smelled 'ugly'. This kind of bias seems to be one of the factors that led some investigators (e.g. Stein et al., 1958) to conclude that young children do not display adult-like, hedonic reactions to odors. However, in a recent study by Schmidt and

Beauchamp (1988), in which odor preferences were assessed using a forced-choice paradigm embedded in a game that focused children's attention directly on the odor stimuli and minimized memory requirements, odor preferences and aversions among 3-year-olds were found to be highly similar to those of adults.

A forced-choice, paired-comparison technique avoids the particular response bias described above but is open to others, such as the consistent choice of either the first or second stimulus in each pair. In order to unambiguously identify such a pattern of response, all possible pairs in a given stimulus series must be presented twice, with the order of presentation of stimuli within each pair counterbalanced. This requirement, however, severely limits the number of stimuli that can be contrasted, as the task quickly becomes too long and tedious to maintain the attention and interest of young children (e.g. contrasting three stimuli requires presentation of six pairs, contrasting four requires 12 pairs and contrasting five requires 20 pairs).

Using the traditional, paired-comparison procedure and a three-stimulus series, we found that preschoolers tend to prefer higher levels of salt in soup than is typical of adults (Cowart & Beauchamp, 1986*b*); in fact, the majority of children tested indicated a preference for the most heavily salted soup included in the stimulus series (0·34M NaCl). That result left open the question of peak salt preference in this age group (i.e. the point beyond which increasing the concentration of salt would make the soup less rather than more pleasant); therefore, studies were initiated to assess young children's responses to even higher levels of salt in soup.

Pilot work indicated that many children refused to complete a stimulus series containing more than four stimuli (or 12 pairs). A series including unsalted soup and soups containing 0·18, 0·34 and 0·56M NaCl was then presented to preschoolers ($N = 14$) recruited from among those having a regular checkup at the Well Baby Clinic of Pennsylvania Hospital (this population is almost exclusively black and of lower socioeconomic status). As shown in Table 2, results from this first, four-stimulus series suggested that a sufficiently high concentration of salt had still not been included to enable observation of a decline in preference among these preschoolers. Consequently, a second series was tested in another 14 children from the same population; that series included unsalted soup and soups containing 0·18, 0·56 and 1·8M NaCl, and the results are also shown in Table 2. As can be seen, the highest salt concentration in this series was preferred by

TABLE 2
Preferred level of salt in soup among preschoolers: number (percentage) of children choosing unsalted and each of three levels of NaCl in two different concentration series

	Plain	0·18M	0·32M	0·56M	1·18M
Series 1 (N = 14)	2 (14%)	3 (22%)	2 (14%)	7 (50%)	Not presented
Series 2 (N = 14)	6 (43%)	1 (7%)	Not presented	5 (36%)	2 (14%)

a smaller percentage of children than the next lower one, although half of the children still chose extraordinarily high concentrations of salt ($\geqslant 0$·56M) as their most preferred. Surprisingly, however, when presented with this series, almost half of the children seemed to exhibit an unusually low salt preference, choosing the unsalted soup as their most preferred.

In effect, repeated presentation of a stimulus containing a level of salt well above that preferred by many of the children tested seemed to lead to the rejection of any amount of salt by a substantial number of them. We also observed this phenomenon in a second, smaller study in which the same children were presented two different stimulus series, only one of which included a highly concentrated salt stimulus (1·0M NaCl) (unpublished data). Although the small sample sizes preclude strong conclusions, these findings suggest that, just as young children's response choices are particularly susceptible to other sources of bias, they may be particularly susceptible to contextual biases.

As reviewed by Booth *et al.* (1983), adult responses to sensory stimuli have also been shown to be distorted, although perhaps not as dramatically, by the presentation of stimuli outside the assessor's tolerated range. These authors developed a scaling procedure for assessing preferred levels of a taste that avoided this problem by allowing for the selection of stimulus levels for each subject such that they were centered around the subject's 'ideal' level. We have adapted that idea to the paired-comparison format.

In the new procedure, children are initially presented with a stimulus pair comprising solutions that contain easily discriminable concentrations of NaCl drawn from the middle of the range of concentrations frequently used in studies of adult, suprathreshold salt perception (specifically, 0·18 and 0·56M NaCl). They are asked to taste each solution and indicate which they like better, without instruction regarding how the solutions differ. Each subsequent pair is then determined by the child's preceding preference choice, and the procedure continues until the child has consecutively chosen a given concentration of salt when it was paired with both a higher and a lower concentration (or has chosen either the unsalted or most concentrated salt stimulus twice in a row). This typically requires presentation of 3-5 pairs. The task is then repeated with the members of the stimulus pairs presented in reverse order (i.e. if the stronger stimulus in the first pair were initially presented first, the weaker one would be presented first in the second series). The order in which stronger and weaker stimuli are presented prevents a child from reaching criterion responding if he or she chooses on the basis of a first or second position bias. The geometric mean of the salt concentrations chosen in the two trial series provides the estimate of the child's most preferred level of salt. Examples of data produced by two young children using this procedure are shown in Fig. 1.

By 'tracking' preferred tastant concentrations in this way, it is possible to assess preference across a wide range of concentrations while at the same time limiting both the number of stimulus pairs that must be presented to each child and the extent to which any child is exposed to tastes he or she finds aversive. As noted, the procedure also guards against consistent position biases. Finally, the two trial runs provide information concerning the consistency of the child's response choices.

The tracking procedure does not yield the extremely skewed or bimodal distributions of salt preferences that were sometimes encountered with traditional, paired-comparison series and that seemed counter-intuitive and likely to be artifactual. Results obtained with the new procedure do, however, continue to indicate that preschoolers are likely to prefer higher levels of salt than are preferred by adults. Moreover, this phenomenon does not appear to be restricted to children from the black, lower socioeconomic population on whom our studies have focused (and who may be exposed to relatively high sodium diets).

Table 3 depicts the distributions of preferred levels of salt in soup obtained from black preschoolers drawn from the Well Baby Clinic ($N = 24$), caucasian preschoolers drawn from middle to upper-middle class families ($N = 28$), and black adults who were the parents of children visiting the Well Baby Clinic ($N = 34$). For the purposes of chi-square analyses, preferred concentrations of salt are characterized as being typical of those previously reported for adults ($\leqslant 0.18$M NaCl;

Early Development of Taste Perception

Order			NaCl Concentration				
	0 M	.1 M	.18 M	.32 M	.56 M	1.0 M	1.8 M
1	___	___	_X_	___	_[X]_	___	___
1	___	___	___	_[X]_	_X_	___	___
1	___	___	_X_	_[X]_	___	___	___
1	___	___	___	___	___	___	___
2	___	___	___	___	___	___	___
2	___	___	___	___	___	___	___
2	___	___	___	___	___	___	___
2	___	___	_[X]_	___	_X_	___	___
2	___	___	_[X]_	_X_	___	___	___
2	___	_X_	_[X]_	___	___	___	___
2	___	___	___	___	___	___	___
1	___	___	___	___	___	___	___
1	___	___	___	___	___	___	___
1	___	___	___	___	___	___	___

Results
Series 1, Most Preferred: 0.32M; Series 2, Most Preferred: 0.18M
Geometric Mean Preferred Concentration: 0.24M

Order			NaCl Concentration				
	0 M	.1 M	.18 M	.32 M	.56 M	1.0 M	1.8 M
2	___	___	_X_	___	_[X]_	___	___
2	___	___	___	_X_	_[X]_	___	___
2	___	___	___	___	_X_	_[X]_	___
2	___	___	___	___	___	_[X]_	_X_
1	___	___	___	___	___	___	___
1	___	___	___	___	___	___	___
1	___	___	___	___	___	___	___
1	___	___	_[X]_	___	_X_	___	___
1	___	___	_X_	_[X]_	___	___	___
1	___	___	___	_X_	_[X]_	___	___
1	___	___	___	___	_X_	_[X]_	___
2	___	___	___	___	___	_[X]_	_X_
2	___	___	___	___	___	___	___
2	___	___	___	___	___	___	___

Results
Series 1, Most Preferred: 1.0M; Series 2, Most Preferred: 1.0M
Geometric Mean Preferred Concentration: 1.0M

FIG. 1. Response patterns of two children in a paired-comparison preference task using the 'tracking' format. The salt was dissolved in a soup broth. X denotes the two stimuli presented in each trial, with the brackets indicating which one the child preferred. In Order 1, the weaker salt stimulus of the pair was presented first; in Order 2, the stronger salt was presented first.

TABLE 3
Preferred level of salt in soup among black and caucasian preschoolers and black adults: number (percentage) of subjects choosing low, moderate and high NaCl concentrations

	$\leqslant 0.18M$	$0.24-0.42M$	$\geqslant 0.56M$
Black children ($N = 24$)	11 (46%)	8 (33%)	5 (21%)
Caucasian children ($N = 28$)	14 (50%)	9 (32%)	5 (18%)
Black adults ($N = 34$)	26 (76%)	7 (21%)	1 (3%)

see, for example, Bertino et al., 1982), moderately high (0·24–0·42M) or extremely high ($\geqslant 0.56M$). The distributions of preferred concentrations for the two groups of preschoolers do not differ ($\chi^2 = 0.08$; $p > 0.25$); however, the distribution of the children's salt preferences (combining the two groups) does differ from that of the adults ($\chi^2 = 8.11$, $p < 0.025$), that is, the children were more likely to prefer the moderately and, especially, the extremely salty soups than were the adults.

Summary
The understanding of human taste responses during late infancy and early childhood is still quite limited. There are several lines of evidence indicating that familiarity/dietary experience begins to impact on taste and food preferences early in life. At the same time, however, young children often prefer more concentrated sweet and salty tastes than they are likely to encounter in their normal diet. Further exploration of relationships between dietary history and taste preferences in early childhood are required. Studies of age-related changes in preferences for both sweet and salty tastes, and how they relate to each other, might also provide insight into the bases for the extreme preferences observed in some young children.

Methodological innovation is greatly needed in this area of research. As illustrated in the foregoing presentation, the types of measures that have been used in studies of older infants and young children are limited and may, in some cases, not be entirely appropriate. In developmental studies, it is particularly important for researchers to be aware of the potential cognitive, communicative, affective and other nonsensory factors that may affect task performance and to try multiple approaches to the same questions. Only in this way can we begin to understand the meaning of children's responses to chemical stimuli.

ACKNOWLEDGEMENTS

The preparation of this paper and research reported herein were supported by NIH Grant RO1 HL31736. We thank the Campbell Soup Company and the staff and clients of the Well Baby Clinic of Pennsylvania Hospital for their assistance. We also thank K. Burgess-Agee, N. Capitillo and K. Rudolph for their technical assistance.

REFERENCES

Abrahams, H., Krakauer, D. & Dallenbach, K. M. (1937). Gustatory adaptation to salt. *American Journal of Psychology,* **49,** 462–9.
Bartoshuk, L. M. (1975). Taste mixtures: Is suppression related to compression? *Physiology and Behavior,* **14,** 643–9.
Beauchamp, G. K. & Cowart, B. J. (1987). Development of sweet taste. In *Sweetness,* ed. J. Dobbing. Springer-Verlag, London, pp. 127–40.
Beauchamp, G. K. & Moran, M. (1982). Dietary experience and sweet taste preferences in human infants. *Appetite,* **3,** 139–52.
Beauchamp, G. K. & Moran, M. (1985). Acceptance of sweet and salty tastes in 2-year-old children. *Appetite,* **5,** 291–305.
Beauchamp, G. K., Cowart, B. J. & Moran, M. (1986). Developmental changes in salt acceptability in human infants. *Developmental Psychobiology,* **19,** 75–83.
Bertino, M., Beauchamp, G. K. & Engelman, K. (1982). Long-term reduction in dietary sodium alters the taste of salt. *American Journal of Clinical Nutrition,* **36,** 1134–44.
Birch, L. L. (1979*a*). Dimensions of preschool children's food preferences. *Journal of Nutrition Education,* **11,** 77–80.
Birch, L. L. (1979*b*). Preschool children's food preferences and consumption patterns. *Journal of Nutrition Education,* **11,** 189–92.
Birch, L. L. & Marlin, D. W. (1982). I don't like it; I never tried it: Effects of exposure on two-year-old children's food preferences. *Appetite,* **3,** 353–60.
Booth, D. A., Thompson, A. & Shahedian, B. (1983). A robust, brief measure of an individual's most preferred level of salt in an ordinary foodstuff. *Appetite,* **4,** 301–12.
Cowart, B. J. (1981). Development of taste perception in humans: Sensitivity and preference throughout the life span. *Psychological Bulletin,* **90,** 43–73.
Cowart, B. J. & Beauchamp, G. K. (1986*a*). Factors affecting acceptance of salt by human infants and children. In *Interaction of the Chemical Senses with Nutrition,* ed. M. R. Kare & J. G. Brand. Academic Press, New York, pp. 25–44.
Cowart, B. J. & Beauchamp, G. K. (1986*b*). The importance of sensory context in young children's acceptance of salty tastes. *Child Development,* **57,** 1034–9.
Crook, C. K. (1977). Modification of the temporal organization of neonatal sucking by taste stimulation. In *Taste and Development: The Genesis of Sweet*

Preference, ed. J. M. Weiffenbach. US Government Printing Office, Washington, D. C., pp. 146-60.
Crook, C. K. (1978). Taste perception in the newborn infant. *Infant Behavior and Development*, **1**, 52-69.
Crook, C. K. & Lipsitt, L. P. (1976). Neonatal nutritive sucking: Effects of taste stimulation upon sucking and heart rate. *Child Development*, **47**, 518-22.
Desor, J. A., Maller, O. & Turner, R. (1973). Taste in acceptance of sugars by human infants. *Journal of Comparative and Physiological Psychology*, **84**, 496-501.
Desor, J. A., Maller, O. & Andrews, K. (1975). Ingestive responses of human newborns to salty, sour and bitter stimuli. *Journal of Comparative and Physiological Psychology*, **89**, 966-970.
Desor, J. A., Maller, O. & Greene, L. S. (1977). Preference for sweet in humans: Infants, children and adults. In *Taste and Development: The Genesis of Sweet Preference*, ed. J. M. Weiffenbach. US Government Printing Office, Washington D.C., pp. 161-72.
Ekman, P. & Freisen, W. V. (1978). *Manual for the Facial Action Coding System*. Consulting Psychologists Press, Palo Alto, CA.
Engen, T. (1974). Method and theory in the study of odor preferences. In *Human Response to Environmental Odors*, ed. J. Johnston, D. Moulton & A. Turk. Academic Press, New York, pp. 121-41.
Engen, T. & Lipsitt, L. P. (1965). Decrement and recovery of responses to olfactory stimuli in the human neonate. *Journal of Comparative and Physiological Psychology*, **59**, 312-16.
Engen, T., Lipsitt, L. P. & Kaye, H. (1963). Olfactory responses and adaptation in the human neonate. *Journal of Comparative and Physiological Psychology*, **56**, 73-7.
Ferrel, M. F., Mistretta, C. M. & Bradley, R. M. (1981). Development of chorda tympani taste responses in the rat. *Journal of Comparative Neurology*, **198**, 37-44.
Filer, L. J. (1978). Studies of taste preference in infancy and childhood. *Pediatric Basics*, **12**, 5-9.
Ganchrow, J. R., Steiner, J. E & Munif, D. (1983). Neonatal facial expressions in response to different qualities and intensities of gustatory stimuli. *Infant Behavior and Development*, **6**, 473-84.
Green, B. G. & Gelhard, B. (1989). Salt as an oral irritant. *Chemical Senses*, **14**, 259-71.
Harris, G. & Booth, D. A. (1987). Infants' preference for salt in food: Its dependence upon recent dietary experience. *Journal of Reproductive and Infant Psychology*, **5**, 97-104.
Hill, D. L. & Almli, C. R. (1980). Ontogeny of chorda tympani responses to gustatory stimuli in the rat. *Brain Research*, **197**, 27-38.
Hill, D. L., Mistretta, C. M. & Bradley, R. M. (1982). Developmental changes in taste response characteristics of rat single chorda tympani fibers. *Journal of Neuroscience*, **2**, 782-90.
Kawamura, Y., Okamoto, J. & Funakoshi, M. (1968). A role of oral afferents in aversion to taste solutions. *Physiology and Behavior*, **3**, 537-42.

Lipsitt, L. P. (1977). Taste in human neonates: Its effect on sucking and heart rate. In *Taste and Development: The Genesis of Sweet Preference,* ed. J. M. Weiffenbach. US Government Printing Office, Washington, D. C., pp. 125-42.

Lipsitt, L. P., Engen, T. & Kaye, H. (1963). Developmental changes in the olfactory threshold of the neonate. *Child Development,* **34,** 371-6.

Maller, O. & Desor, J. A. (1973). Effects of taste on ingestion by human newborns. *Fourth Symposium on Oral Sensation and Perception: Development in the Fetus and Infant,* ed. J. F. Bosma. US Government Printing Office, Washington, D. C., pp. 279-91.

McBurney, D. H., Smith, D. V. & Shick, T. R. (1972). Gustatory cross adaptation: Sourness and bitterness. *Perception & Psychophysics,* **11,** 228-32.

Mistretta, C. M. & Bradley, R. M. (1983). Neural basis of developing salt taste sensation: Response changes in fetal, postnatal, and adult sheep. *Journal of Comparative Neurology,* **215,** 199-210.

Peterson, F. & Rainey, L. H. (1910). The beginnings of mind in the newborn. *Bulletin of the Lying-in Hospital City of New York,* **7,** 99-102.

Preyer, W. (1901). *The Mind of the Child* (H. W. Brown, trans.). Appleton, New York. (Originally published 1882).

Rosenstein, D. & Oster, H. (1988). Differential facial responses to four basic tastes in newborns. *Child Development,* **59,** 1555-68.

Rovee, C. K. (1972). Olfactory cross-adaptation and facilitation in human neonates. *Journal of Experimental Child Psychology,* **13,** 368-81.

Schmidt, H. J. & Beauchamp, G. K. (1988). Adult-like odor preferences and aversions in 3-year-old children. *Child Development,* **59,** 1136-43.

Shinn, M. W. (1900). *The Biography of a Baby.* Houghton Mifflin, New York.

Stein, M., Ottenberg, P. & Roulet, N. (1958). A study of the development of olfactory preferences. *American Medical Association Archives of Neurological Psychiatry,* **80,** 264-6.

Steiner, J. E. (1973). The gustofacial response: Observation on normal and anencephalic newborn infants. *Fourth Symposium on Oral Sensation and Perception: Development in the Fetus and Infant,* ed. J. F. Bosma. US Government Printing Office, Washington, D. C., pp. 254-78.

Steiner, J. E. (1977). Facial expressions of the neonate infant indicating the hedonics of food-related chemical stimuli. In *Taste and Development: The Genesis of Sweet Preference,* ed. J. M. Weiffenbach. US Government Printing Office, Washington, D. C., pp. 173-89.

Steiner, J. E. (1979). Human facial expression in response to taste and smell stimulation. In *Advances in Child Development and Behavior,* Vol. 13, ed. H. Reese & L. P. Lipsitt. Academic Press, New York, pp. 257-95.

Weiffenbach, J. M. (1977). Sensory mechanisms of the newborn's tongue. *Taste and Development: The Genesis of Sweet Preference,* ed. J. M. Weiffenbach. US Government Printing Office, Washington, D. C., pp. 205-16.

Weiffenbach, J. M. & Thach, B. T. (1973). Elicited tongue movements: Touch and taste in the newborn human. In *Fourth Symposium on Oral Sensation and Perception: Development in the Fetus and Infant,* ed. J. F. Bosma. US Government Printing Office, Washington, D. C., pp. 232-44.

2

Effects of Aging on Sensory Functioning: Implications for Dietary Selection

CLAIRE MURPHY
Department of Psychology, San Diego State University, San Diego, California 92182-0551, USA

&

MAGDALENA M. GILMORE
Department of Psychology, Brown University, Providence, Rhode Island 02912, USA

INTRODUCTION

The chemical senses (taste, smell and trigeminal sensitivity) play important roles in preparing the body for food and signalling information about its nature and palatability. Although the chemical senses function in concert in the perception of food flavor, each makes a unique and independent contribution to that perception. The sense of taste alerts the brain to the presence of sweet, sour, bitter and salty substances in the oral cavity. The sense of smell processes the volatiles which produce the subtleties and complexities of food flavors. Odors can reach the olfactory receptors both through the nose and through the mouth. The importance of the olfactory component to food flavor can be dramatically illustrated by blocking the nostrils before ingestion. In many cases, it will be impossible to identify the food or beverage. The trigeminal sense provides the brain with information about other sensations in the oral and nasal cavities: warmth, cold, irritation, pungency. The heat of hot peppers, the coolness of menthol and the bite of horseradish are signalled by the trigeminal nerve.

It is becoming increasingly clear that aging affects the functioning of the chemical senses in significant ways. This chapter briefly reviews some of the literature which demonstrates these effects and highlights the results of studies with significant emphasis on chemosensory preference. A more complete understanding of alterations in taste and olfaction and, in particular, chemosensory preference in the elderly may provide some insight into dietary selection by the elderly. Since compromised nutritional status is a problem for significant numbers of elderly people, dietary selection may be particularly important for the health and well-being of the elderly.

TASTE

Threshold
Many investigators have demonstrated that the taste system shows modest increases in threshold with age (Richter & Campbell, 1940; Harris & Kalmus, 1949; Bouliere et al., 1958; Hinchcliffe, 1958; Cooper, et al., 1959; Balogh & Lelkes, 1961; Kalmus & Trotter, 1962; Glanville, et al., 1964; Kaplan et al., 1965; Smith & Davies, 1973; Grzegorczyk, et al., 1979; Murphy, 1979; Schiffman, et al., 1979; Dye & Koziatek, 1981; Schiffman, et al., 1981; Moore, et al., 1982; Weiffenbach, et al., 1982; Gilmore & Murphy, 1989; and see Murphy, 1979 and 1986 for reviews). Some qualities are affected more than others. The magnitude of the threshold increase and its role in perception of real-world stimuli are currently under debate.

Suprathreshold intensity
Using magnitude estimation and magnitude matching, several investigators have demonstrated age-related changes in suprathreshold taste intensity perception (See Murphy, 1986, for a more complete review). Enns et al., (1979) found no alteration in the slope of the psychophysical function for taste from young adulthood to old age, although adults had flatter slopes than children. Bartoshuk (1983) has reported stability of slopes for taste function in old age, with the exception of some flattening near threshold for bitter, which she attributed to lack of dental hygiene. Schiffman and Clark (1980), Schiffman et al. (1981), Hyde and Feller (1981), and Cowart (1983) all reported some flattening of the slopes of taste functions in elderly subjects, suggesting some decline in the ability of the elderly to track

increases in stimulus concentration. Weiffenbach *et al.*, (1986) reported poorer performance by elderly subjects in a taste magnitude estimation task, based on their analysis of intra-class correlations.

Age-related changes in taste functioning at the suprathreshold level have been demonstrated to be more quality-specific than those at the threshold level. The greatest age-related losses appear to be associated with the perception of bitterness and the least with the perception of sweetness (Weiffenbach *et al.*, 1986; Murphy & Gilmore, 1989; Cowart, in press).

For example, Murphy and Gilmore (1989) designed a study to examine age-associated suprathreshold taste loss. Young and elderly subjects used magnitude matching (Marks & Stevens, 1980) to rate the perceived intensity of the single tastants sucrose, caffeine, sodium chloride and citric acid. Perceived intensity for individual tastants in the following two-component mixtures was also assessed: sucrose/caffeine, sucrose/citric acid, and sucrose/sodium chloride. For the first set of mixtures, the 16 stimuli were aqueous solutions of sucrose (0·0, 0·15, 0·30, 0·60M), caffeine (0·0, 0·0025, 0·005 and 0·01M), all possible mixtures of the two solutes, and deionized water. In the second and third sets of mixtures, citric acid (0·0, 0·0015, 0·003, 0·006M) and NaCl (0·0, 0·10, 0·20, 0·40M) replaced the caffeine. Pilot work showed these stimulus ranges to be of approximately equal subjective intensity for college students. Weights (20, 50, 100, 200, 400, 750 g) were used as the calibration continuum.

The elderly subjects' intensity judgments for bitter were significantly lower than those of the young subjects for both unmixed and mixed components. A smaller, but significant, effect of age on intensity for citric acid was also demonstrated when unmixed components were compared. Only sucrose/caffeine mixtures showed significant age-related differences in the relative contributions of the two components to the perceived intensity of the mixtures. Sucrose and NaCl showed no significant effects of age on slopes of psychophysical functions or on average perceived intensity (Murphy & Gilmore, 1989). Because of the existence of different perceptual contexts for young and elderly subjects, and the potential influence of context on magnitude matches, the possibility exists that age effects on NaCl and sucrose intensity are underestimated. It is clear, however, that any such effects will be smaller than the effect of aging on bitter perception.

The results of this study support the hypothesis that aging in the taste system is quality specific, and they are consistent with electrophysiological data on taste responses in the chorda tympani in rats (McBride & Mistretta, 1986). Neither the psychophysical data nor the animal electrophysiological data rule out central effects, but both suggest age effects at the periphery. Research on human taste papillae suggests that many papillae are responsive to more than one taste quality (Cardello, 1981) and that the mechanism for specificity, if it exists, must be at the lower level of the taste bud or the taste cell. There are conflicting reports regarding loss of taste papillae, buds and cells in aged humans (Schiffman, 1986). Differential age effects on human taste qualities suggest that membrane and receptor function may play more of a role in age-related loss than do decreased numbers of taste papillae. Thus, it may be more fruitful to investigate the functional status of individual taste buds and cells.

Discrimination thresholds
A study by Gilmore and Murphy (1989) was designed to simplify the process of assessing taste sensitivity in elderly subjects. Some methodologies for assessing taste sensitivity (e.g. the scaling techniques) can be problematic when testing elderly individuals. Thus, a forced-choice task in which the subject's only requirement was to choose the strongest of two stimuli was chosen. By manipulating concentration levels and utilizing statistical analyses an individual's just noticeable difference (JND), an indicator of how much of a physical stimulus is needed in order to obtain a perceptible difference in taste sensation, was determined. The ratio of the JND to a particular standard concentration is the Weber Ratio (WR). The average WR for taste has been determined to be 0·20 (Pfaffmann et al., 1971), hence one must increase the stimulus 20% before a subject will perceive a noticeable difference in taste sensation. The study investigated

(1) whether elderly individuals require a greater increase in the physical stimulus in order to get a perceptible difference in sensation than do young subjects, and
(2) whether the size of the WR is dependent upon not only age, but also the stimulus tested.

Since the results of a previous study (Murphy & Gilmore, 1989) resulted in greater age-related differences in the perception of bitter stimuli than in the perception of sweet stimuli, these two tastants were examined in the present study.

Weber Ratios for 12 elderly females 67–77 years of age and 12 young females 18–25 years of age were determined using the method of constant stimuli. The standard concentrations were 0·15, 0·30 and 0·60M sucrose and 0·0025, 0·005 and 0·01M caffeine. Six comparison stimuli were used with each standard, three more concentrated than the standard and three less concentrated than the standard, in increments of 0·12 (McBride, 1983). Each subject also performed a recognition threshold task.

The authors were, therefore, able to be sure that in the discrimination task subjects were tasting stimuli which were above the recognition threshold.

The results of this study once again support the hypothesis of quality specificity in the aging taste system. WRs generated by the elderly subjects were significantly larger than those generated by the young subjects for the two highest concentrations of caffeine, but not the low concentration. WR's did not differ between the two age groups for sweet stimuli. Using the 0·005M concentration of caffeine as an example, it took a 74% increase in caffeine to elicit a perceptible difference in the intensity of bitterness for the elderly subjects, whereas it took only a 34% increase in caffeine to elicit a perceptible difference in the intensity of bitterness for the young subjects.

Determining discrimination thresholds is a simple task for elderly subjects and yields information about taste sensitivity that the scaling techniques do not. As a supplement to other testing techniques, WRs may help us understand why elderly subjects do not perceive bitter stimuli in the same manner that young subjects do, as well as help us understand more about why the perception of sweet stimuli appears to be more stable over the lifespan. It may be useful to determine WRs as an index of taste sensitivity when testing individuals with impaired cognitive status, such as Alzheimer's Disease patients, or with individuals who have impairments in other sensory systems and thus would ordinarily be prevented from performing magnitude-matching tasks.

The overall enjoyment of eating may be affected by the quality-specific taste losses associated with aging. In general, the ability to perceive bitterness is diminished in elderly people. This tastant serves as the underlying component in numerous foods and in particular many beverages. Bitter substances such as herbs and spices are sometimes added to foods to enhance their overall flavor. Perhaps selecting a diet which compensates for this sensory loss may improve the palatability of foods for some elderly people.

OLFACTION

Threshold
When food is placed in the mouth, a person perceives taste information (e.g. information about sweet, sour, bitter and salty), and also olfactory and trigeminal information from the myriad of volatiles in foods and beverages. These volatiles travel retronasally to stimulate the olfactory system and thus add additional chemosensory stimulation (Murphy et al., 1977). Age-related olfactory loss may thus play a major role in flavor perception for the elderly. Olfactory threshold studies have yielded almost unanimous agreement that there is age-related decline, both for stimuli which are largely olfactory and for stimuli which are largely trigeminal (Chalke & Dewhurst, 1957; Fordyce, 1961; Joyner, 1963; Kimbrell & Furtchgott, 1963; Minz, 1968; Venstrom & Amoore, 1968; Strauss, 1970; Schiffman et al., 1976; Murphy, 1983; Murphy et al., 1985; and see Murphy, 1986, for a review). Thus, components of the overall flavor complex may be missing from the elderly person's perception of a food flavor.

Nasal airway resistance
The authors have been interested in the possibility that changes in nasal airflow contribute to the increase in olfactory threshold associated with aging. In a recent study, both nasal airway resistance and the extent of allergic rhinitis and bacterial infection in the elderly and the young, were assessed, in order to determine their influence on olfactory threshold changes which might otherwise be ascribed to aging, per se. A summary of these findings can be found in Murphy et al., (1985). A more detailed account of this study and a more extensive follow-up study will be found in Murphy et al. (submitted). Forty-eight people were studied. Of these, half were young ($M = 21$ years) and half were elderly ($M = 73$ years). All were active, community-dwelling persons who reported good to excellent health and no hospitalizations in the preceding year.

Olfactory threshold for butanol was tested using a two-alternative, forced-choice, staircase procedure. Each time a subject made an error concentration was increased. When the subject was correct, the same concentration was presented again, to a criteria of four correct in a row. Nasal airway resistance was calculated from simultaneous measurements of nasal inspiratory flow-rates and pressures utilizing the active, uninasal, anterior, rhinomanometric technique described in Jalowayski *et al.* (1983). Nasal mucosal samples from the mid-inferior portion of the inferior turbinate were taken using a disposable, plastic rhinoprobe. These samples were examined for the presence of cells indicative of allergic rhinitis, bacterial infection, and upper respiratory viral infection.

Analysis of co-variance confirmed the presence of an age effect ($p < 0.0001$) on olfactory threshold in addition to an effect of nasal airway resistance ($p < 0.05$). Contrary to what is seen in nasal disease, nasal airway resistance decreased in the elderly relative to the young. Hence, a simple explanation of increased nasal obstruction cannot be evoked in an effort to describe the relationship between nasal airway resistance and olfactory threshold in the elderly. There was very little incidence of allergy or infection in either age group.

Suprathreshold intensity

The authors' studies using magnitude estimation as well as studies by other investigators using both magnitude estimation and magnitude matching have investigated the effects of aging on suprathreshold intensity scaling of odor (Stevens *et al.*, 1982; Murphy, 1983; Stevens *et al.*, 1984; Stevens & Cain, 1987). These studies have demonstrated even greater age effects on olfactory perception than on taste perception. In some cases the projected intercept of the psychophysical function was increased in the elderly, in others the slope of the function was decreased: that is, increasing concentration increased perceived intensity at a greater rate for young subjects than for elderly subjects.

The implications of these results for perception of odors in the environment, as well as of foods and beverages, are clear: older people perceive odors as less intense than young people do. Hence, odors, either alone or in combinations (as they are in food flavors), will be less strong for the elderly than for the young. Those formulating food specifically for the elderly, or broadly enough to span all age groups in a graying population, should be aware of this information.

Odor identification

Another clue to the influence of sensory loss on dietary selection lies in studies of odor identification in the elderly. Schiffman (1979) reported that the elderly had poorer identification of blended foods. Murphy (1981, 1985) confirmed that the elderly's ability to identify foods in the mouth was significantly compromised, and further demonstrated that the difficulty lies more with the olfactory system than with the taste system. Older people had great difficulty identifying blended foods on the basis of odor.

Other studies of the elderly's ability to identify odors also indicate impairment, whether the odors are sniffed from jars or from microencapsulated labels (Schemper et al., 1981; Doty et al., 1984; Murphy & Cain, 1986).

CHEMOSENSORY PREFERENCE

Given substantial evidence suggesting both threshold and suprathreshold age-related differences in sensitivity of the chemical senses, the question arises: how is preference for chemosensory stimuli affected by age?

Laird and Breen (1939) reported an increased preference for tart taste over sweet taste in older subjects and to a lesser degree in females of all ages. Enns et al. (1979) measured preferences for sucrose in 21 fifth graders, 27 college students, and 12 elderly subjects, and found that fifth graders and the elderly females showed a lesser preference for sweeter sucrose solutions than did the college age participants and males of all ages. Dye & Koziatek (1981) measured pleasantness of suprathreshold aqueous solutions of sucrose for 79 diabetic and nondiabetic veterans. When pleasantness judgments of men 41–65 years were compared with those of men 65–88 years, the three-way interaction of age, patient group, and sucrose level was significant at the 0·0001 level. Older nondiabetic subjects increased pleasantness ratings as concentration increased over the range 0·125–1M. Younger nondiabetic subjects rated 0·25M as the pleasantest stimulus in the series and decreased their pleasantness ratings with further increases in concentration.

Although Desor et al., (1975) demonstrated that, in a sample of 618 children and 140 adults, 9–15 year olds preferred greater sweetness and saltiness than did adults 18–64 years old, no additional age effects emerged when preferences of adults 18–29 years and 45–64 years were compared. The proportions of the adult group falling into these age ranges were not reported. Age-related preference differences in the amount of salt added to chicken broth were found by Pangborn et al., (1983). Older (36–66-year-old) subjects added more salt than younger (17–32-year-old) subjects when allowed to salt to preference.

Age-related shifts in preference for some odors were documented by Moncrieff (1966). Mere exposure to odors (Cain & Johnson, 1978) or to olfactory-taste mixtures, presented orally, can produce shifts in pleasantness (Murphy, 1982). Murphy (1982) also demonstrated effects of context on the pleasantness of chemosensory stimuli. Exposure-related pleasantness shifts occurring over a lifetime could result in altered food and odor preferences in the elderly. Preference shifts need not necessarily be correlated with an underlying alteration in the sensory systems. However, a change in the ability of a chemosensory system to process intensity information, reflected in a change in slope or up-down position of the psychophysical functions, would imply a change in preferred concentration and hedonic judgments, since intensity has been shown to be a powerful predictor of hedonic tone (Moskowitz et al., 1976). If, for example, the salty function flattened with age, then a stimulus which is too salty for a young person will be less salty for an elderly person and may move from negative to positive in hedonic tone.

Two useful measures of hedonics are the peak preferred concentration (i.e that one concentration in a series which is chosen as the most preferred) and the quantitative pleasantness judgement. Both measures may be important in assessing age-associated changes in chemosensory hedonics. For example, in a series of four concentrations of sucrose, young and elderly subjects might both choose a given concentration as the most preferred. However, the elderly might rate the next and highest concentration as pleasant while the young subjects might rate the sweetest stimulus as unpleasant. For this reason the study described below was designed to yield information regarding the existence of age-associated differences in both of these measures.

Pleasantness of sucrose and sodium chloride
A large-scale study of taste preference across the adult lifespan was recently conducted (Murphy & Withee, 1986). This study was designed to investigate the existence of age-associated changes in preference for various concentrations of tastants in aqueous solution and of the same tastants in complex chemosensory mixtures. Complete details regarding this study can be found in Murphy and Withee (1986). Three questions were addressed: first, for the stimuli salt, sugar or citric acid, does the most preferred concentration in a series differ in different age groups? Second, are there age-associated changes in pleasantness

judgments for various concentrations of salt, sugar or citric acid? Third, does the background in which a stimulus is presented significantly affect its hedonic tone?

One hundred young, 100 middle-aged, and 100 older adults rated, on bipolar line scales, the pleasantness of sucrose, citric acid and NaCl, each presented in four concentrations in deionized water; the same four concentrations were also presented in appropriate beverage bases.

Analysis of Variance (ANOVA) on the pleasantness ratings (Murphy, 1982) showed significant effects of age, background, stimulus and concentration. Mean pleasantness ratings were less negative for elderly participants than for either young or middle-aged participants. Stimuli were judged less pleasant overall in an aqueous base than in a beverage base, and concentration significantly affected ratings. The significant interactions of age with stimulus, background with stimulus, and of age, stimulus and concentration, provide interesting insights into age-associated differences in pleasantness. Young and middle-aged participants found salt significantly less pleasant than did elderly subjects. Similarly, the mean pleasantness ratings for sucrose were significantly higher for older subjects than for middle-aged, but not for young subjects (as seen in Fig. 1). The two highest concentrations of sucrose were also rated as pleasanter by the elderly participants than by the younger participants. Citric acid was less pleasant than NaCl for elderly subjects, but the reverse was true for young and middle-aged subjects.

Pleasantness judgments of all three stimuli were significantly affected by the background base in which they were presented. Sucrose and NaCl were both rated more pleasant in the beverage base, but the background produced greater differences in the pleasantness of NaCl than in the pleasantness of sucrose. Salt was rated as pleasanter by the elderly than by other participants regardless of its background (see Fig. 2). When presented in deionized water, increasing NaCl concentration drove pleasantness down for all age groups. However, when presented in beverage base, the two middle salt concentrations were preferred to the lowest by both middle-aged and elderly raters. Citric acid was perceived as less pleasant in the beverage base than in the aqueous base.

The interaction of age with stimulus and concentration was statistically significant. While pleasantness ratings generally decreased with increasing concentration for both NaCl and citric acid, the ratings for sucrose increased for the first three concentrations, then decreased

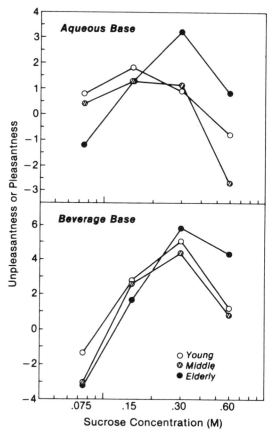

Fig. 1. Pleasantness or unpleasantness ratings for sucrose in water and in beverage base. From Murphy and Withee, 1986. Copyright (1986) by the American Psychological Association. Reprinted by permission of the publisher.

at the fourth to produce inverted U-shaped functions. Elderly subjects rated the two highest concentrations of sucrose as pleasanter than the lowest concentration, and they rated them pleasanter than the younger subjects did.

The data were analyzed for peak preferred concentration by a three-factor (age, background, stimulus) ANOVA with repeated measures. Results showed that the peak preferred concentration differed as a

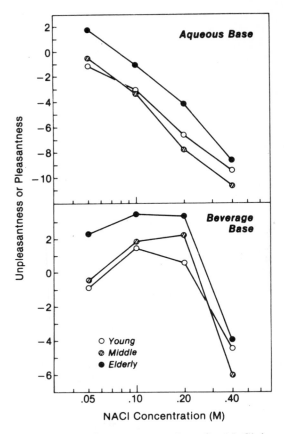

Fig. 2. Pleasantness or unpleasantness ratings for NaCl in water and in beverage base. From Murphy and Withee, 1986. Copyright (1986) by the American Psychological Association. Reprinted by permission of the publisher.

function of background and of stimulus. The mean preferred concentration was higher in beverage base than in aqueous solution. Regardless of background, participants preferred higher concentrations of sucrose than of NaCl or citric acid. This relatively gross measure did not capture the age effect seen in the analysis of the individual pleasantness ratings.

These findings may be important in understanding dietary selection in elderly people, particularly those on low-salt or sugar-free diets.

Preference for high-salt foods and preference for sugar or items containing large amounts of sugar may have negative health consequences since they may result in dietary selections which have significant negative effects on the older person. As a group, the elderly have increased incidence of hypertension and diabetes. There is also a gradual rise in systolic blood pressure with age which may have clinical sequelae which are as yet not completely understood. There is also a tendency for blood sugar to rise with age and the clinical implications of this are not entirely clear. Diet and exercise have been shown to be beneficial in controlling blood pressure and blood glucose levels (Brownell, 1982; Martin & Dubbert, 1982; Lindgarde et al., 1983). Compliance to diet regimes can be difficult at any age, but may be particularly difficult for elderly persons. Decreased energy expenditure results in decreased caloric needs and therefore reduced intake is necessary to maintain energy balance.

Sucrose and NaCl in high concentrations were rated as pleasanter by elderly than by younger participants. Although other explanations cannot be ruled out, the most obvious possible explanation for this effect is sensory. Older people may, for example, rate higher concentrations of salt as pleasanter simply because these concentrations are less salty to them than to younger subjects who generally rate very high concentrations of salt as unpleasant. This sensory hypothesis follows from studies demonstrating some loss of suprathreshold intensity with aging for some of the simple tastants (Murphy & Gilmore, 1989; Cowart, in press) and for amino acids (Schiffman & Clark, 1980). The results of the present study suggest that an experiment designed to directly test the ability of intensity to predict chemosensory hedonics across the lifespan would be worthwhile.

As with any cross-sectional aging study, the question of cohort differences in the present study arises. Environmental influences may have interacted with sensory influences on perception of flavor. The significant age effects on pleasantness ratings in the present study suggest the importance of future longitudinal studies.

The fact that pleasantness judgments differ when the taste stimuli are in beverage base versus aqueous base underscores the importance of other chemosensory elements in determining pleasantness. Age-associated changes in olfactory perception may be partially responsible for the differences in pleasantness judgments made by the young, middle-aged and elderly subjects in these experiments. For example, an older person might judge a stronger concentration of salt in the beverage base as more pleasant, not necessarily because he desired more salty taste, but because he desired an overall stronger flavor. He or she could compensate for reduced sensory input from volatiles by increasing sensory input to the taste system. Experiments considering the ability of elderly and young subjects to identify blended foods, with and without the sense of smell, clearly demonstrated that the olfactory system was more affected by the aging process than the taste system (Murphy, 1981, 1985). Others have since come to the same conclusion (Stevens et al., 1984).

This study also emphasized the importance of the diluent in chemosensory preference research. Research investigating pleasantness of stimuli in aqueous solution may not present a clear picture of real-world preferences of humans. Salt in water and sugar in water are not generally consumed by adults. On the average, subjects in the present study chose as their most preferred a concentration of NaCl in vegetable juice which exceeded their preferred concentration in deionized water. This result seems to be related to subjects' clear preference for salt in foods and beverages naturally consumed, in spite of their distaste for concentrated solutions of salt in water. In the past, many studies of taste preference have been conducted with deionized water or distilled water as the diluent. The generalizability of results with pure tastants such as NaCl and sucrose in water to real world preference for salt and sugar is questionable.

Elderly people rated salt and sugar pleasanter at higher concentrations than younger subjects did. The reasons for this may be cultural, contextural or sensory. Simultaneous investigation of chemosensory preference and chemosensory intensity in a single group of elderly and young persons would help elucidate the effects of the aging process on chemosensory perception and pleasantness.

NUTRITION AND CHEMOSENSORY PERCEPTION

Many elderly people have less than optimal nutritional status. Nutritional deficiencies may result from decreased nutrient intake (Beauchene & Davis, 1979), or from lowered rates of absorption and utilization in the elderly (Yearick et al., 1980). Up to 41% of elderly participants show deficient levels of serum protein, and 20% show deficiency in serum albumin, according to Yearick et al., (1980) and Jansen and Harrill (1977).

The research described below was designed to investigate the relationship between nutritional status and chemosensory preference in elderly persons. A complete description of these experiments may be found in Murphy and Withee (1987). Nutritional status was operationally defined as the biochemical indices of total protein, albumin, and blood urea nitrogen (BUN).

The effects of aging and biochemical status on preference for casein hydrolysate were first investigated. The hypotheses were two: first, that older participants would rate high concentrations of the amino acid mixture as more pleasant than young participants would; second, that participants with lower biochemical status would prefer higher concentrations of casein hydrolysate than would those with higher biochemical status.

The participants were ten persons 18–26 years old ($M = 22.8$) and 16 persons 70–92 years of age ($M = 84.0$). The chemosensory stimuli consisted of an amino acid-deficient soup base to which were added the following concentrations of casein hydrolysate: 0, 1, 2, 3, 4 and 5% w/v. Stimuli were prepared in deionized water, refrigerated, and then heated before presentation. The participants rated pleasantness of the chemosensory stimuli using a bipolar line scale. Blood was drawn from each person, usually on the same day, and an independent laboratory provided assays of serum total protein, albumin, and BUN. On the average, elderly participants had lower protein and albumin, and higher BUN values.

An analysis of variance was performed to examine the effect of age on peak preferred concentration (PPC), which was defined as the concentration (0, 1, 2, 3, 4 or 5% casein hydrolysate) most preferred by each participant. Results showed that the elderly preferred higher concentrations of casein hydrolysate ($M = 3.0\%$) than did the young ($M = 0.5\%$). Similar analyses examining the effect of the three blood measures (grouped above and below the median) on PPC indicated that higher concentrations of casein hydrolysate were preferred by those with higher values of BUN, and those with lower serum albumin. There was no difference in preferred concentration associated with serum protein level.

These results suggested, in a small sample, the association of age and biochemical status in the perceived pleasantness of casein hydrolysate. A second experiment was designed to further investigate these variables as well as to determine the effect of perceived intensity on preference for casein hydrolysate. There were three hypotheses: (1) that older participants would prefer higher concentrations of the amino acid mixture, (2) that participants of lower biochemical status would prefer higher concentrations of casein hydrolysate, and (3) that perceived intensity would be predictive of preference.

Of the 40 participants in the second study, 20 were 18–26 years of age ($M = 20\cdot15$) and 20 were 65–83 years of age ($M = 70\cdot75$). The method of magnitude matching (Stevens & Marks, 1980) was used to determine intensity of the same six chemosensory stimuli used in the first experiment as well as of six auditory stimuli which were included only for the purposes of matching. Participants also rated pleasantness of both auditory and chemosensory stimuli using the bipolar line scale described above. Biochemical status was determined as in the first experiment.

All three biochemical measures showed significant age effects. Compared to the young, elderly participants showed lower levels of serum protein and albumin, and elevated levels of BUN. For the remaining analyses a composite index of the three blood measures was created by combining each subject's standard (z) scores (protein + albumin − BUN).

Flavor intensity judgments were normalized by auditory intensity judgments (Stevens *et al.*, 1982) before being subjected to ANOVA. Both age group and concentration significantly affected intensity estimates. Older participants made lower intensity estimates for amino acids (geometric mean = 3·3) than did young participants (geometric mean = 5·4). All participants successfully tracked increases in concentration. Age group differences in intensity were similar at all concentrations of casein hydrolysate: there were no differences in slopes of the psychophysical functions. Biochemical status showed no significant relationship to perceived intensity.

In order to test in an ANOVA the effects of age, biochemical status and perceived intensity on the preferred concentration of casein hydrolysate (PPC), a geometric mean intensity rating was computed for each participant across the six concentrations. ANOVA showed that age and blood status (as measured by the biochemical index described above) were significantly related to PPC, but perceived intensity was not. Elderly participants preferred higher concentrations of casein hydrolysate ($M = 1\cdot4\%$) then did young participants ($M = 0\cdot4\%$). Across age, participants with lower composite biochemical indices also preferred higher concentrations of amino acids ($M = 1\cdot5\%$) than participants with higher biochemical status ($M = 0\cdot4\%$). The percentages of each age group and within each level of biochemical status who preferred each of the concentrations of casein hydrolysate showed essentially the same pattern as was found in the first experiment. The present studies thus revealed differences in preferences for casein hydrolysate, a nutritionally significant chemosensory stimulus. These differences were significantly dependent upon both nutritional status, assessed biochemically, and age. The elderly participants' higher preferred concentration of casein hydrolysate was not simply due to generally lower perceived intensity. This result does not completely rule out the possibility that a chemosensory deficit underlies the elderly's preference for higher concentrations. If flavor components of casein hydrolysate fall below an elderly person's odor or taste threshold, then the overall flavor may be significantly altered and its acceptability changed, independent of overall perceived intensity.

The present research indicates that the interrelationship between chemosensory perception and nutrition in the geriatric population is an area which warrants further study. The authors plan to explore this interrelationship in a series of further studies. They have already begun to investigate whether dietary intake, assessed by three-day dietary records, will be related to chemosensory preference in the same way that biochemical status is. They hope to begin to understand the role that altered chemosensory function in the elderly may play in the compromised nutritional status seen in a significant proportion of the elderly population.

ACKNOWLEDGEMENTS

Supported by NIH Grant No. AG04085 from the National Institute on Aging. Magdalena M. Gilmore (formerly Jensen) is currently at Brown University, Providence, Rhode Island. We are grateful to Elizabeth Konowal, Carol Randall, Michele J. Reed, R. Blair Skinner, Jill M. Sniffen and Jeanne Withee for excellent technical assistance.

REFERENCES

Balogh, K. & Lelkes, K. (1961). The tongue in old age. *Gerontologia Clinica*, **3**, Suppl. 38–54.
Bartoshuk, L. M. (1983). Effects of aging on chemical senses. Paper presented at the Fifth Annual Meeting of the Association for Chemoreception Sciences, Sarasota FL, May.

Beauchene, R. E. & Davis, T. A. (1979). The nutritional status of the aged in the U.S.A. *Age,* **2**, 23.

Bouliere, F., Cendron, H. & Rapaport, A. (1958). Modification avec l'age des seuils gustatifs de perception et de reconnaissance aux saveurs salee et sucre, chez l'homme. *Gerontologia,* **2**, 104–12.

Brownell, K. D. (1982). Obesity: understanding and treating a serious, prevalent, and refractory disorder. *J. Consult. & Clin. Psychol,* **50**, 820–40.

Cain, W. S. & Johnson, F. (1978). Lability of odor pleasantness: Influence of mere exposure. *Perception,* **1**, 459–65.

Cardello, A. V. (1981). Comparison of taste qualities elicited by tactile, electrical and chemical stimulation of single human taste papillae. *Percep. Psychophys.,* **29**, 163–9.

Chalke, H. D. & Dewhurst, J. R. (1957). Accidental coal-gas poisoning. *Brit. Med. J.,* **2**, 915–17.

Cooper, R. M., Bilash, I. & Zubeck, J. P. (1959). The effect of age on taste sensitivity. *J Geront.,* **14**, 56–8.

Cowart, B. J. (1983). Direct scaling of the intensity of basic tastes: A life span study. Fifth Annual Meeting of the Association for Chemoreception Sciences, Sarasota, Fl., April.

Cowart, B. J. (1989). Relationships between taste and smell across the adult life span. In *Nutrition and the Chemical Senses in Aging,* ed. C. Murphy, W. S. Cain and D. M. Hegsted. New York Academy of Sciences, New York.

Desor, J. A., Green, L. S. & Maller, O. (1975). Preferences for sweet and salty tastes in 9 to 15-year old and adult humans. *Science,* **190**, 686–7.

Doty, R. L., Shaman, P., Applebaum, S. L., Giberson, R., Siksorski, L. & Rosenberg, L (1984). Smell identification ability: Changes with age. *Science,* **226**, 1441–3.

Dye, C. J. & Koziatek, D. A. (1981). Age and diabetes effects on threshold and hedonic perception of sucrose solutions. *J. Geront.,* **36**, 310–15.

Enns, M. P., Van Itallie, T. B. & Grinker, J. A. (1979). Contributions of age, sex and degree of fatness on preferences and magnitude estimation for sucrose in humans. *Physiol. Behav.,* **22**, 999–1003.

Fordyce, I. D. (1961). Olfaction tests. *Brit. J. Ind. Med.,* **18**, 213–15.

Gilmore, M. M. & Murphy, C. (1989). Aging is associated with increased Weber ratios for caffeine, but not for sucrose. *Percep. Psychophys.,* **46**, (6) 555–9.

Glanville, E. V., Kaplan, A. R. & Fischer, R. (1964). Age, sex and taste sensitivity. *J. Geront.,* **19**, 474–8.

Grzegorczyk, P. B., Jones, S. W. & Mistretta, C. M. (1979). Age-related differences in salt taste acuity. *J. Geront.,* **34**, 834–40.

Harris, H. & Kalmus, H. (1949). The measurement of taste sensitivity to phenylthiourea (PTC). *Ann. Hum. Genet., (London),* **15**, 24–31.

Hinchcliffe, R. (1958). Clinical quantitative gustometry. *Acta Oto-laryngol.,* **49**, 453–66.

Hyde, R. J. & Feller, R. P. (1981). Age and sex effects on taste of sucrose, NaCl, citric acid and caffeine. *Neurobiol. Aging,* **2**, 315–18.

Jalowayski, A. A., Yuh, Y., Kozilo, J. A. & Davidson, T. A. (1983). Surgery for nasal obstruction — evaluation by rhinomanometry. *Laryngoscope,* **93**, 341–5.

Jansen, C. & Harrill, I. (1977). Intakes and serum levels of protein and iron for 70 elderly women. *Am. J. Clin. Nutrit.*, **30**, 1414–22.
Joyner, R. E. (1963). Olfactory acuity in an industrial population. *J. Occup. Med.*, **5**, 37–42.
Kalmus, H. & Trotter, W. R. (1962). Direct assessment of the effect of age on PTC sensitivity. *Ann. Hum. Genet, (London)*, **26**, 145–9.
Kaplan, A., Glanville, E. & Fischer, R. (1965). Cumulative effect of age and smoking on taste sensitivity in males and females. *J. Geront.*, **20**, 334–7.
Kimbrell, G. M. & Furchtgott, E. (1963). The effect of aging on olfactory threshold. *J. Geront.*, **18**, 364–5.
Laird, D. A. & Breen, W. J. (1939). Sex and age alterations in taste preferences. *J. Am. Diet. Assoc.*, **15**, 549–50.
Lindgarde, F., Malmquist, J. & Balke, B. (1983). Physical fitness, insulin secretion, and glucose tolerance in healthy males and mild Type 2 diabetes. *Acta Diabet. Latina*, **20**, 33–40.
Marks, L. E. & Stevens, J. A. (1980). Measuring sensation in the aged. In *Aging in the 1980's: Psychological Issues*, ed. L. W. Poon, American Psychological Association, Washington, D. C., 592–8.
Martin, J. E. & Dubbert, P. M. (1982). Exercise applications and promotion in behavioral medicine: Current status and future direction. *J. Consult. Clin. Psych.*, **50**, 1004–7.
McBride, R. L. (1983). A JND-scale/category scale convergence in taste. *Percep. Psychophys.*, **32**, 77–83.
McBride, M. R. & Mistretta, C. M. (1986) Taste responses from the chorda tympani nerve in young and old fischer rats. *J. Geront.*, **41**, 306–14.
Minz, A. I. (1968). Condition of the nervous system in old men. *Zeitschrift fur Alternsforschung*, **21**(3), 271–7.
Moncrieff, R. W. (1966). *Odour Preferences*. John Wiley, New York.
Moore, L. M., Neilson, C. R. & Mistretta, C. M. (1982). Sucrose taste thresholds: Age-related differences. *J. Geront.*, **37**, 64–9.
Moskowitz, H. R., Kumraiah, V., Sharma, K. N., Jacobs, H. L. & Sharma, S. D. (1976). Effects of hunger, satiety and glucose load upon taste intensity and taste hedonics. *Physiol. & Beh.*, **16**, 471–5.
Murphy, C. (1979). The effects of age on taste sensitivity. In *Special Senses in Aging*, ed. S. S. Han and D. H. Coons. University of Michigan Institute of Gerontology, Ann Arbor, Michigan, pp. 21–33.
Murphy, C. (1981). Effects of aging on chemosensory perception of blended foods. Third Annual Meeting of the Association for Chemoreception Sciences, Sarasota, Fl., April.
Murphy, C. (1982). Effects of exposure and context on hedonics of olfactory-taste mixtures. In *Selected Sensory Methods: Problems and Approaches to Measuring Hedonics*. ed. J. T. Kuznicki, R. A. Johnson and A. F. Rutkiewic. ASTM STP 773, 1982, American Society for Testing and Materials, Philadelphia, PA, pp. 60–70.
Murphy, C. (1983). Age-related effects on the threshold, psychophysical function, and pleasantness of menthol. *J. Geront.*, **38**, 217–22.
Murphy, C. (1985). Cognitive and chemosensory influences on age-related changes in the ability to identify blended foods. *J Geront.*, **40**, 47–52.

Murphy, C. (1986). Taste and smell in the elderly. In *Clinical Measurement of Taste and Smell*, ed. H. L. Meiselman and R. S. Rivlin. Macmillan, New York, pp. 343–71.
Murphy, C. & Cain, W. S. (1986). Odor identification: The blind are better. *Physiol. Behav.*, **37**, 177–80.
Murphy, C. & Withee, J. (1986). Age-related differences in the pleasantness of chemosensory stimuli. *Psych. Aging*, **1**, 312–18.
Murphy, C. & Withee, J. (1987). Age and biochemical status predict preference for casein hydrolysate. *J. Geront.*, **42**, 73–7.
Murphy, C. & Gilmore, M. M. (1989). Quality-specific effects of aging on the human taste system. *Percep. Psychophys.* **45**, 121–8.
Murphy, C., Cain, W. S. & Bartoshuk, L. M. (1977). Mutual action of taste and olfaction. *Sensory Processes*, **1**, 204–11.
Murphy, C., Nunez, K., Withee, J. & Jalowayski, A. A. (1985). The effects of age, nasal airway resistance and nasal cytology on olfactory threshold for butanol. *Chem. Sens.*, **10**, 418.
Murphy, C., Gilmore, M. M., Jalowayski, A. A., Davidson, T. M. & Nunez, K. (1990). Olfactory threshold, nasal airway resistance, and nasal pathology in young and elderly persons.. *Chemical Senses* (submitted).
Pangborn, R. M., Braddock, K. S. & Stone, L. J. (1983). Ad Lib. Mixing to Preference vs Hedonic Scaling: Salts in Broths and Sucrose in Lemonade. AChemS V Poster Presentation, Sarasota, Fl.
Pfaffmann, C., Bartoshuk, L. M. & McBurney, D. H. (1971). Taste psychophysics. In *Handbook of Sensory Physiology* vol. IV, part 2, Gustation. ed. L. M. Beidler. New York, Springer-Verlag.
Richter, C. & Campbell, K. (1940). Sucrose taste thresholds of rats and humans. *Am. J. Physiol.*, **128**, 291–7.
Schemper, T., Voss, S. & Cain, W. S. (1981). Odor identification in young and elderly persons: sensory and cognitive limitations. *J. Geront.*, **18**, 446–52.
Schiffman, S. S. (1977). Food recognition by the elderly. *J. Geront.*, **32**, 586–92.
Schiffman, S. S. (1979). Changes in taste and smell with age: Psychophysical aspects. In *Sensory Systems and Communication in the Elderly*, ed. J. M. Ordy and K. R. Brizzee. Raven Press, New York, pp. 227–46.
Schiffman, S. S. (1986). Age-related changes in taste and smell and their possible causes. In *Clinical Measurement of Taste and Smell*. ed. H. L. Meiselman and R. S. Rivlin. Macmillan, New York, pp. 326–42.
Schiffman, S. S. & Clark, T. B. (1980). Magnitude estimates of amino acids for young and elderly subjects. *Neurobiol. Aging*, **1**, 81–91.
Schiffman, S. S., Moss, J. & Erickson, R. P. (1976). Thresholds of food odors in the elderly. *Exp. Aging Res.*, **2**, 389–98.
Schiffman, S. S., Hornak, K. & Reilly, D. (1979). Increased taste thresholds of amino acids with age. *Am. J. Clin. Nutr.*, **32**, 1622–7.
Schiffman, S. S., Lindley, M. G., Clark, T. B. & Makins, C. (1981). Molecular mechanism of sweet taste: Relationship of hydrogen bonding to taste sensitivity in both young and elderly. *Neurobiol. Aging*, **2**, 173–85.
Smith, S. E. & Davies, P. D. (1973). Quinine taste thresholds: A family study and a twin study. *Ann. Hum. Genet. (London)*, **37**, 227–32.

Stevens, J. C. & Marks, L. E. (1980). Cross-modality matching functions generated by magnitude estimation. *Percep. Psychophys.,* **27**, 379–89.

Stevens, J. C. & Cain, W. S. (1987). Old-age deficits in the sense of smell as gauged by thresholds, magnitude matching, and odor identification. *Psychol. Aging,* **2**, 36–42.

Stevens, J. C., Plantinga, A. & Cain, W. S. (1982). Reduction of odor and nasal pungency associated with aging. *Neurobiol. Aging,* **3**, 125–32.

Stevens, J. C., Bartoshuk, L. M. & Cain, W. S. (1984). Chemical senses and aging: Taste versus smell. *Chem. Senses,* **9**, 167–79.

Strauss, E. L. (1970). A study on olfactory acuity. *Ann. Otol. Rhinol. Laryngol.,* **79**, 95–104.

Venstrom, D. & Amoore, J. E. (1968). Olfactory threshold in relation to age, sex or smoking. *J. Food Sci.,* **33**, 264–5.

Weiffenbach, J. M., Baum, B. J. & Burghauser, R. (1982). Taste thresholds: quality specific variation with human aging. *J. Geront.,* **37**, 700–6.

Weiffenbach, J. M., Cowart, B. J. & Baum, B. J. (1986). Taste intensity perception in aging. *J. Geront.,* **41**, 460–8.

Yearick, E. S., Wang, M. L. & Pisias, S. J. (1980). Nutritional status of the elderly: Dietary and biochemical findings. *J. Geront.,* **35**, 663–71.

3

The Perception of Complex Taste Stimuli

JAN H. A. KROEZE
Laboratory of Psychology, University of Utrecht, PO Box 80.140, 3508TC, Utrecht, The Netherlands

WHAT DOES TASTE CONTRIBUTE TO FOOD AND DRINK?

The living body has a physiological need for a group of chemicals in order to preserve its structural and functional integrity over time. However, people do not eat primarily because they acknowledge this technical fact. They do not analyse food into its amino acid and carbohydrate components, neither do they calculate the amounts of water needed to dissolve and emulsify the consumed substances. Most of the time people simply eat and drink because they like it. However, food intake is not completely driven by hedonic motives. Opinions and beliefs play a part too. Apart from chemicals, people also 'digest' information. The media supply information about how the body functions, how food is metabolized, which substances may be potential health hazards and what role vitamins and additives might play. This information makes many consumers 'food conscious'. They no longer exclusively select and eat because they like it, but beliefs related to health and longevity contribute as well. Taste only contributes partially to the total percept of an item of food. In many cases the smell of food is more important than taste. The feel of food in the mouth related to texture as well as temperature contribute to the complex percept. Even food colours, well known to be important for selection and choice, may contribute to perceived taste (Pangborn, 1967; Johnson & Clydesdale, 1982; Hyman, 1983). Johnson *et al.* (1982) showed that in a cherry-flavoured beverage part of the sweetness can be replaced by colour. Red, in particular, appeared to be a small but significant sweet substitute. Because there is more to eating than taste, experiments in which smell, touch, temperature, and other attributes are excluded do not provide results that can be applied in a straightforward way to everyday eating behaviour. Laboratory experiments, some of which are presented in this chapter, are designed to analyse sensory or perceptual mechanisms.

In the literature a distinction is made between hedonic and intensity scaling. For instance, when a subject is asked to express in a number the degree of liking of a series of sodium chloride intensities and of a series of sucrose intensities, two different functions result. On the other hand when subjective intensity is scaled sodium chloride and sucrose have about equal positive power functions. When subjects evaluate stimuli on a hedonic dimension they generally have a synthetic attitude, i.e. they do not primarily analyse the stimulus cognitively into its components. They produce a unidimensional like-dislike score, even with a complex mixture (Coombs, 1979). This is not to say that subjects are unable to analyse a taste into the intensities of its sensational components if asked to do so. But it requires a certain effort to ignore the like-dislike dimension during intensity scaling.

PERCEPTUAL INTEGRATION AND ANALYSIS

Complexity of a stimulus and of a sensation are often confused. When asked to count the number of different tastes, subjects do not necessarily report the number of stimulus components. They report, or at least try to report, the number of components in the percept. These may or may not coincide with the stimulus components. Relatively complex stimuli may often give rise to a simple sensation. This is because the senses primarily integrate separate pieces of sensory input into a percept. Everyday perception is object-oriented and introspective analysis of perception into separate sensations requires conscious attention or even training (McBurney, 1986). For example, separate line features may be integrated into a single pattern as in face recognition. Integration is more or less the rule in perception. Even very complex percepts with parts originating from different senses may occur and give an impression of unity. Such a composition is called a *pattern*.

The perceiver has no immediate cognitive access to the integrative process itself. His awareness is not even required. The sensory system has been built to perform this act of integration on the raw sensory input.

Often more than one meaning is attached to the word 'analysis'. For example, it is stated that audition is an analytical sensory system and that taste is both analytic and synthetic (Erickson, 1978, 1982; Erickson & Covey, 1980). Two sound waves (e.g. respectively 2000 and 4000 Hz) reaching the inner ear cause maximum vibration on different spots of the basilar membrane. Each of these spots is connected to a distinct group of neurons in the auditory nerve. Thus the auditory system is built in such a way that the two sound waves are received at spatially separate 'entrances'.

Higher in the auditory pathway this separation, or 'tonotopic' organization as it is called, is preserved. Apparently, hardly any effort is needed to distinguish two tones when the corresponding physical frequencies are sufficiently separated. They are not even mixed so there is no need to analyse them. In as much as analysis refers to the process of dissecting a complex sensation into two or more separate parts, one cannot say that the human ear performs analysis. When, however, the initial sorting of frequencies by the inner ear is defined as analysis then analysis there is. It is specifically those two definitions, the initial sorting of frequencies and the subsequent separation of sensational or perceptual attributes, that are often confused.

Gestalt psychology emphasized the wholeness of a percept (Köhler, 1947). Thus the drawing of a square does not present itself to the observer as a set of four lines. To the observer it is primarily a *square*. The perceptual system has combined the separate physical elements into a meaningful pattern. And this percept has priority with respect to the parts. But if instructed to do so, subjects can easily analyse this percept in terms of lines, corners, and distances. Should we declare pattern vision synthetic because of this gestalt priority, or analytic because of the subject's ability to analyse the percept into its parts?

Food perception provides us with a similar example: a certain pattern of tastes, smells and texture may be recognized immediately as cheese, but on request subjects can analyse it into several taste, smell and texture components. Analysis implies attentional shifts between parts of a percept. Between shifts, unidimensional psychophysical judgments may be made by the subject. Analysis of a percept requires more effort if it is more complicated (e. g. a lot of parts), more tightly organized (i. e. the total impression of wholeness is so strong that the subjects can hardly direct their attention to a part of it), or more peripherally synthesized (as with yellow light, where subjects are not able to analyse it into relative contributions of a red and a green sensitive system). Subjects have no cognitive access to the peripheral colour coding mechanism, so no matter how much effort they invest in the attentional analysis of yellow into its components, they will never succeed.

But this is nothing special. Most of the integrative activities going on in the organism are out of reach for the attentional mechanism. Conceiving of analysis and synthesis as opposites on a continuum may easily lead to problems when the same term ('analysis') is used for initial sensory sorting as well as for conscious separation of perceptual attributes. Different types of physical energy as converted by receptors may be combined peripherally (at or near the receptor) or centrally (eventually within reach of the attentional mechanism).

Let us consider audition once more. When a mixture of four different tones is played, the listener can easily tell them apart. Therefore, according to Erickson (1982), hearing may be called an analytical sense. One might put it another way and state that the tones were not even integrated, so no analysis was needed or even possible. Why then call a sensory system 'analytical'? The author would agree, of course, if audition were designated as largely non-integrative or non-synthetic. This name then would tell us that the auditory system does not integrate different wavelengths into one or a smaller number of tones. The misunderstanding is in viewing synthesis (integration) and analysis as the two poles of a unidimensional concept. In common-sense language it may be phrased as follows: integration is what *perception* does and analysis is what *you* do. Analysis is a conscious, time-consuming activity, i.e. you know that you are doing it. Integration may be carried out early in the flow of information or late. It may even occur step-wise: partially integrated parts may be integrated further at a higher (later) level in the sensory system (process). Most of the time you are unaware of how and by what mechanism patterns are created.

Analysis and attention
Analysis of a complex stimulus is accompanied by alternating selective attention. It is more difficult to attend to stimulus components separately if they are more spatiotemporally integrated. Kahneman (1973) pointed out that, besides selectivity, attention may be characterized by effort. Analysis of a complex stimulus (i.e. focusing attention on one attribute according to some instruction) may or may not require

effort as indicated by brain evoked potentials. When subjects are offered a deviant tone (1240 Hz, occurring on only 10% of occasions) in a sequence of identical tones (1000 Hz), evoked potentials indicate that no attentional effort is required for discrimination (Sams *et al.*, 1984). The negligible effort required for pitch discrimination, as reflected by the P300 wave in the EEG, was also noticed by Polich (1987). Cognitive analysis may be quick, as in the case of distinguishing yellow and red in orange, or impossible, as in the case of analysing white light.

Sensory analysis
The conscious effort to identify and judge different sensational components in an object, be it a piece of food, a soft drink or a perfume, is called *sensory analysis*. The components may be scaled and a multidimensional profile representing the perceptual pattern may be constructed. People can be trained to adopt the analytical attitude and to use the judgment procedures required to analyse percepts. Such results say more about the structure of the percept than about the physical structure of the perceived object. Sensory analysis data can only tell us about a physicochemical property or component of a stimulus if, on the basis of independent experimental data, a unique and unambiguous relationship has been established beforehand between a physical property of the stimulus and a certain perceptual experience. If that sensation is reliably reported on a subsequent occasion by members of a sensory panel it may be assumed that the corresponding physical property of the stimulus is present.

Simple and complex stimuli
It is assumed tacitly that the term taste mixture means a physical mixture of two or more distinct substances. Thus when sodium chloride and sucrose are physically mixed the result is a mixed taste stimulus. So mixing two taste stimuli is *procedurally* similar to mixing two wavelengths of light. Procedural similarity between a light mixture and a mixture of chemicals is to a certain extent superficial. For example, we know from light that after analysing its components with a prism, no further analysis of the light as a stimulus is possible. But a sucrose stimulus, widely recognized as inducing a single sweet sensation, can easily be split up into two independent substances each having a (sweet) taste (glucose and fructose). This can also be done with Quinine Hydrochloride, a purely bitter substance. It can be split up into two independent substances: quinine, tasting bitter, and HCl, tasting sour.

Numerous other examples can be found to demonstrate the fact that, from a physicochemical point of view, there are no basic taste stimuli analogous to basic colour stimuli.

However, a better comparison may be possible if we assume that in the processing sequence a step present in one sense may be absent in another sense. Thus, the first steps in vision — separate adsorption of groups of wavelengths and photochemical transitions — are absent in taste. Unlike taste, vision has to bridge a distance. If we insist on making comparisons between vision and taste, the best equivalent starting point may be the step in which, in both senses, a chemical stimulus affects a receptor cell.

Studying mechanisms versus eating and drinking
Eating and drinking are complex behaviours, in which many bodily systems are involved. Taste researchers, being aware of this complexity, have great difficulty in understanding its complex natural form. This is one of the reasons why the contributing mechanisms are studied in isolation. With this purpose in mind, mechanisms such as adaptation, mixture suppression, habituation, detection, recognition, contrast, cross enhancement which are all involved in a complex interplay, are studied separately in the laboratory. Well-circumscribed stimuli rather than real life food products are used in artificial stimulation procedures. In one case the tongue is put into a flow chamber and exposed to a controlled stimulus flow, in another case pieces of filter paper are placed on the tongue and only a restricted quantifiable response is allowed. These procedures are not primarily meant to predict consumer behaviour, but to improve understanding of the sensory systems related to eating and drinking.

ADAPTATION

Subjects report fading of sensation when a stimulus is applied uninterruptedly to the same location of the tongue (for instance, when a subject is asked to report at regular intervals the saltiness of a continuous sodium chloride stimulus, the type of result in Fig. 1 is obtained). This is adaptation.

Sensory adaptation certainly plays a role in the sensation during food intake. However, the undisturbed adaptation phenomenon as observed under restricted laboratory conditions is not as clear cut in everyday

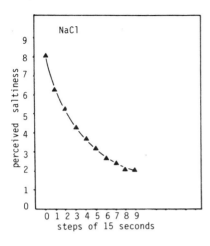

FIG. 1. Saltiness of a filter paper (diameter 10 mm) soaked in a 0·32 mol/litre NaCl solution) estimated every 15 s (Ganzevles & Kroeze, 1987b).

circumstances. Restricted experimental conditions make it possible to produce complete taste adaptation (Gent & McBurney, 1978; Ganzevles & Kroeze, 1987a). It is important to avoid mouth and tongue movements to demonstrate adaptation. Abrahams et al. (1937) noticed that, with some subjects in the final phase of adaptation (when the sensation had already disappeared), sudden recovery phenomena occurred, notwithstanding the still-flowing taste solution. It appeared that the subjects could not perfectly fulfill the requirement not to move the tongue during the stimulation period. The investigators asked one subject — before he had completely adapted — to move his tongue. Even slight movements prevented complete adaptation. Von Békésy (1965) found that the irregularities in smooth adaptation curves could be traced back to tongue movements.

Adaptation and eating

During eating, tongue movements which result from chewing and swallowing are essential because they help to avoid complete adaptation. It would be a shame if soon after the beginning of a delicious meal your sensory system adapted completely. Apparently this does not happen. The tongue has thousands of tiny folds and is completely covered on the upper side with (non-tasting) filiform papillae. Taste substances, more or less emulsified and partially dissolved in water (contained in the food) or saliva, are retained in these structures. O'Mahoney (1972a, b) showed that residuals of taste stimuli already expectorated may be very persistent. Thorough and repeated rinsings are required to reduce the levels of NaCl to a degree not different from salt in saliva. As soon as the tongue moves the taste buds may be stimulated or re-stimulated by these residuals. But not only the taste substances are enclosed and freed in the sequential chewing and swallowing movements. Access to taste buds situated below the epethelial surface (fungiform papillae), in folds (foliate papillae) or walls (circum-vallate papillae) varies continuously with movement. Thus, one moment taste cells are stimulated and hence start to adapt and, the next moment, they are sealed off and start to recover. As there are thousands of taste cells on the tongue, an overall equilibrium between recovery and stimulation is established. This is why overall taste varies during a meal but does not adapt completely. Furthermore, during normal eating smell plays an important role (Burdach & Doty, 1987) and subjects, even when paying attention to it, have difficulty in separating taste and smell sensations. Smell sensations originating from food in the mouth are perceived as oral rather than nasal by most subjects. In particular, the loss of smell is often experienced as a loss of taste (Schechter & Henkin, 1974; Mozell et al., 1969; Doty & Kimmelman, 1986).

As the olfactory stimulation caused by mouth movements and swallowing is highly variable, the olfactory system does not adapt strongly to food odours during food-intake, whereas at the same time it may adapt almost completely to uniformly dispersed environmental odours.

Most readers may be familiar with this phenomenon from their own experience. A freshly painted restaurant has a noticeable paint smell immediately after entering it. After a while the uniformly dispersed odour is not noticed any more because of sensory adaptation, but the non-uniformly distributed food odours are still noticed.

Non-uniformly distributed taste substances in a food product may give rise to a unique pattern of partial adaptation equilibria, which gradually changes as chewing proceeds.

There may be a further advantage of partial adaptation. In order to explain this we first look at the results of an experiment by McBurney et al. (1967). These investigators determined differential sensitivity to 0·1M NaCl, expressed as the so-called Weber fraction ($\Delta I/I$),for the unadapted and partially adapted tongue. The Weber fraction is the smallest intensity change (ΔI) of a stimulus (I) that can be perceived by a subject. It is often used as a measure of differential sensitivity and there are several procedures to determine it. Differential sensitivity was 0·20 with the unadapted tongue, but only 0·075 with the tongue adapted. So, after adaptation, differential sensitivity was two-and-a-half times better than before adaptation. One may argue that tasting differences during a meal is at least as welcome as absolute taste. The experiment of McBurney et al. shows that taste adaptation is not merely a sensitivity loss. It looks like an advantageous deal: you lose a bit of absolute sensitivity but in turn get increased differential sensitivity. Furthermore, this phenomenon fits in with a general rule in perception that intensity *change* rather than intensity *per se* is the effective stimulus. Undisturbed or even increased differential sensitivity may be one of the reasons that professional tasters in the food industry have so few problems with adaptation. Also, to them differential sensitivity is of more use than extreme absolute sensitivity. McBurney (1984) citing Keidel et al. (1961) concludes that adaptation is best understood as a mechanism for adjusting the system so that it is maximally sensitive to changes in stimuli near the prevailing intensity. This is not only true for taste, but for most (perhaps even all) sensory systems. It is immediately apparent in vision, where light adaptation (or recovery as in dark adaptation) is a physiological means of adjusting the retina to another level of background illumination.

The locus of adaptation

Zotterman (1971) made electrophysiological recordings of the chorda tympani of human subjects. He observed that the decrease of spike frequency during continuous taste stimulation matched the decrease of reported sensation. On the basis of this correlation he concluded that adaptation is peripherally determined and, in Zotterman's own words, 'there is no need to postulate the existence of a central adaptation mechanism'. Heck and Erickson (1973) ascribe adaptation to occupation rate of receptor sites. With many sites already occupied the adsorption rate is lower than with a few sites occupied. Since, according to their theory, the effective taste stimulus is the *rate* at which receptor sites are occupied, the sensation evoked by a stimulus applied to slightly occupied tissue will be stronger than that evoked by the same stimulus applied at highly occupied tissue.

Cross adaptation

Decreased perceived intensity of a taste stimulus or lowered sensitivity to it after exposure to a chemically different taste stimulus is called cross adaptation. Considerable cross adaptation is found between compounds with similar taste qualities. No, or only minor amounts of, cross adaptation are found between stimuli that have different taste qualities. Self-adaptation (i.e. when exposure and test stimulus are identical) is always stronger than cross adaptation. The amount of cross adaptation between substances may be used to indicate to what degree two stimuli compete for identical receptor sites. If two stimuli do not cross adapt at all, it may be assumed that their respective binding sites are independent. With independent binding sites no peripheral *competition-based* mixture suppression can occur.

MIXTURE SUPPRESSION

Mixture suppression in taste may occur when two or more substances of qualitatively different taste are mixed. Thus, when sucrose and NaCl are mixed, the sweetness and saltiness of the mixture may be less than when judged on the basis of equimolar unmixed components. Our experiments here show that about 20% of human subjects did not show mixture suppression.

There are several reasons for this failure to produce mixture suppression. One is that there is considerable variation between subjects who do show suppression. This may be already a provisional clue to the localization of the suppression mechanism. Sensory mechanisms have the tendency to be more plastic, and to display larger between-subject variation, as their determinants are located higher in the nervous system. The literature on decrease or increase of intensity in mixtures can be summarized as follows. Sucrose sweetness is enhanced by the addition of sub-threshold concentrations of NaCl (Fabian & Blum, 1943; Beebe Center et al., 1959; Kamen et al., 1961; Pangborn, 1962; Indow, 1969). Sub-threshold concentrations of NaCl also increase the sensitivity threshold to sucrose (Anderson, 1955). This enhancing effect of NaCl is mainly due to the sweet taste of NaCl solutions (Renqvist, 1919; Richter & McClean, 1939; Kahn, 1951; Bartoshuk *et al.*, 1964; O'Mahoney *et al.*, 1976; Cardello & Murphy, 1977; Bartoshuk, 1978). Anderson (1955) assumes that the increase in sensitivity to sucrose can be attributed to the general excitatory action of the sodium ion in animal tissue, initially discovered by Loeb (1901). However this general physiological effect is clearly only a small part of the sweetness enhancing effect of NaCl, the major part originating in the sweetness of

NaCl itself. Moderate and strong concentrations of NaCl suppress sucrose sweetness (Beebe Center et al., 1959; Kamen et al., 1961; Pangborn, 1962; Indow, 1969; Bartoshuk, 1975; Kuznicki and Ashbaugh, 1979; Kroeze, 1979, 1982a). Glucose and fructose sweetness are also suppressed by NaCl (Moskowitz, 1972).

Mixture effects in other than sweet–salty mixtures have been reported too. Citric acid suppresses sucrose sweetness (Pangborn, 1960, 1961) and the sweetness of glucose and fructose (Stone et al., 1969; Moskowitz, 1972). The reverse, suppression of citric acid sourness by sucrose has also been shown (Fabian & Blum, 1943; Pangborn, 1960; Kamen et al., 1961). Near-threshold amounts of NaCl suppress citric acid sourness (Fabian & Blum, 1943; Pangborn, 1960; Kamen et al., 1961; Pangborn & Trabue, 1967). This may be due to the side taste of NaCl, since, as mentioned, sweet reduces sour. The effect of stronger concentrations of NaCl was reported to decrease sourness of citric acid in one study (Kamen et al., 1961) and to enhance citric acid sourness in another (Pangborn & Trabue, 1967). Bartoshuk (1975) found that the sourness of HCl was increased slightly when mixed with NaCl. No sourness suppression was found in her study. This puzzling contradiction between acids and NaCl may be solved by assuming that side taste effects may obscure suppression, or, in some cases, even cause enhancement. The experimental evidence for this possibility will be discussed in more detail in the next section.

The general picture emerging from the literature is that suppression may be found with strong stimuli, whereas most cases of enhancement are found with weak, near-threshold stimuli.

Conditions affecting the amount of mixture suppression
Several conditions affect the amount of suppression found in mixture experiments. First and foremost, suppression depends on the type of suppressing taste substance. For instance, sweetness is suppressed significantly more in a sucrose/quinine sulphate mixture than in a sucrose/NaCl mixture, even when the subjective intensities of all compounds in an unmixed state are identical (Kroeze, 1980). The physical intensity of the suppressing component is another important variable: a weak component suppresses the other component in the mixture less than a strong component. A third variable relevant to the amount of suppression is the relative number of unmixed to mixed stimuli in a specific experiment. The greater the number of unmixed stimuli, the greater the amount of mixture suppression. It was shown

(Kroeze, 1982d) that this frequency effect can be predicted by Helson's adaptation-level theory (Helson, 1964). This frequency effect can be neutralized by using equal numbers of unmixed and mixed stimuli in a mixture-suppression experiment.

In daily life the frequencies of different mixtures and unmixed substances as they appear in food and drink vary tremendously between, and considerably within, subjects. This may be one of the determinants of individual differences in mixture suppression outside the laboratory.

Mixture enhancement and side tastes

Although the overall result was suppression in our mixture experiments, some subjects did not show suppression but showed enhancement instead. As already mentioned this appeared true for about 20% of all subjects. Visual inspection of data sheets from the experiments revealed that subjects with enhancement in a binary mixture generally reported high side tastes. For example, subjects displaying sweetness enhancement in a sucrose–NaCl mixture reported higher NaCl-sweetness than subjects showing suppression.

Sweetness of NaCl may be considered a side taste. Side tastes are defined as those taste sensations that are qualitatively distinct from the main taste of a stimulus, and have a lower intensity than the main taste. Side tastes can be assessed by making a taste profile of a stimulus. A profile consists of magnitude estimates on each of the following five dimensions: sweetness, saltiness, sourness, bitterness, and other tastes. The use of less than the categories mentioned results in a partial profile. Other categories may be used as well, but then considerable semantic problems arise, since the taste vocabulary, at least in the western culture, contains only four distinct taste adjectives. Apparently these four are of so overwhelmingly phenomenological importance that no other categorical taste adjectives have emerged. Other adjectives are directly linked to some known food (e.g. strawberry-like) or express a hedonic tone (e.g. disgusting, delicious). The existence of side tastes besides the main taste of a stimulus is a familiar phenomenon and it may be considered the rule rather than the exception. For example, many sugars, besides having a sweet main taste, show a bitter side taste. NaCl in weak concentrations tastes sweet, but, according to Renqvist (1919), beyond a concentration of 0·05M the sweet side taste vanishes.

From the above it may be inferred that, when a subject judges sweetness of NaCl–sucrose mixtures, he in fact judges a combination of sweetness of the sucrose component and the sweet side taste of the NaCl component. Occasionally subjects are found whose sweet taste in the NaCl profile is stronger than the amount of suppression of sucrose-sweetness in the NaCl–sucrose mixture. In such cases more sweetness is added by the suppressing component than is lost by mixture suppression, resulting in net sweetness enhancement. After expressing the amount of specific mixture suppression in a certain mixture (e.g. sweetness suppression in a sucrose–NaCl mixture) as a function of the corresponding side taste of the suppressing component (e.g. NaCl-sweetness), a significant negative relationship emerges that can be equally well described by a linear and a power function.

When, for each subject, the side taste effect is removed, no mixture enhancement is left in the data of our experiments. Enhancement effects in a study of Bartoshuk (1975) also disappeared after removal of side taste effects (Kroeze, 1982a). It may be concluded that one must be careful in assuming that mixture enhancement takes place.

Has mixture suppression a central or peripheral origin?

From mixture suppression studies in which substances are combined and then judged, it was impossible to extract knowledge about the mechanism underlying suppression phenomena. More complex experiments were required to learn more about the location of mixture suppression and its possible mechanism. In one experiment (Kroeze, 1978) subjects estimated the saltiness of $0.3M$ NaCl after adaptation to $0.3M$ NaCl (i.e. self-adaptation) and after adaptation to the mixture of $0.3M$ NaCl and $0.3M$ sucrose. It appeared that sweetness reduction after adaptation to the mixture and after adaptation to unmixed sucrose were not different. In other words, the mixture was as potent as an adapting substance as the unmixed NaCl solution. From control conditions in the same experiment it was clear that strong mixture suppression existed: saltiness of the same mixture that was used as an adapting stimulus was estimated to be significantly lower than saltiness of unmixed NaCl.

The finding that a stimulus reduced in saltiness by mixture suppression has nevertheless unchanged adapting capacity can only be understood if it is assumed that mixture suppression occurs after (is more central than) adaptation.

Regression analysis showed that saltiness after adaptation to sodium chloride and after adaptation to the mixture were perfectly positively correlated. On the other hand, no correlation was found between saltiness of the mixture and saltiness after self-adaptation. This was taken as evidence that adaptation and mixture suppression, apart from having different locations, are independent as well. In other words: for a certain individual you cannot predict the degree of masking if you have his adaptation data and vice versa.

In the experiments we applied only small amounts of stimulus material (0·8 ml/s) to a restricted area of one tongue half. Later, Lawless (1982a) repeated the experiments, but he used the whole tongue and a dorsal flow five times as high as ours. With NaCl-saltiness his results replicated ours, but after adaptation to the NaCl-sucrose mixture, sucrose showed a small but significant extra sweetness as compared to the sweetness after adaptation to itself. Lawless concluded that besides central mixture suppression a small amount of mixture suppression exists peripheral to the site of adaptation. In order to reconcile his results with ours he carried out a second experiment (Lawless, 1982a). In this experiment he adjusted his procedure to ours in two steps, which resulted in three conditions: whole tongue flow of 4 ml/s (as in his first experiment), a 4 ml/s flow to one tongue half, and, thirdly, a reduced flow of 0·8 ml/s to one tongue half. The third condition resembled our procedure most. It appeared that after restriction of the area the peripheral suppression component still showed up, but after flow reduction it had disappeared and the results replicated our earlier results. Thus a high flow rate reveals a small amount of sweetness suppression in the NaCl-sucrose mixture, of which the origin is peripheral to the locus of adaptation. The fact that flow rate rather than area makes the difference suggests that a peri-receptor factor and not a receptor factor may be the cause of this peripheral mixture interaction.

Another way to investigate this problem of peripheral versus central mixture suppression is to adapt a subject to one component of a mixture and then measure the amount of suppression of the other component. In binary mixtures, adapting to one component releases the suppression of the other component (Bartoshuk, 1975; Lawless, 1979, 1982b; Gillan, 1982). Thus the mixture behaves as if it were a single taste substance. It is evident that by this procedure one of the mixture components is removed below the level at which tastes suppress each other.

Lawless (1979) reasoned that if mixture suppression were a neural affair, then suppression release might also be achieved after suppressing the sweet taste of sugar with Gymnema sylvestre (Gymnema sylvestre is a plant extract that removes the sweet taste of sugar). He succeeded in doing this and took it as extra evidence that mixture suppression is of neural origin. The neural pathway is long and complicated and it cannot be deduced, from the results mentioned so far, where in the taste system the locus of mixture suppression is to be found. Adaptation is most probably a receptor process (Zotterman, 1971; Diamant *et al.,* 1965), and thus the suppression mechanism may still be located anywhere beyond the outer periphery. Lawless' Gymnema sylvestre experiment leaves us none the wiser, since the locus of action of this taste modifier is at the receptor level, too.

Habituation and mixture suppression
The suppression release found after adaptation may also be obtained in a different way. It is known that after cessation of the stimulus the sensory system immediately starts to recover from adaptation. Recovery from adaptation is complete within 1 min with most taste substances that easily dissolve in water. This is true for sensitivity as was convincingly show by Hahn (1934). Even with a 15% NaCl solution sensitivity is back to normal after less than 30 s. With stronger suprathreshold stimuli we investigated recovery during the inter-stimulus interval. The results of this investigation are summarized in Fig. 2.

Because taste intensity is completely recovered in so short a time, we may offer subjects repeatedly the same stimulus with inter-stimulus

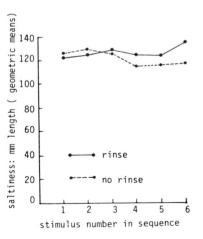

FIG. 2. Mean saltiness estimates of a sequence of 6 identical stimuli (0·194 mol/litre NaCl) replicated 8 times over 9 subjects presented with a capillar flow (0·8 ml/s) at intervals of 60 s. The solid line represents the sequence with distilled-water rinsing during the inter-stimulus interval. The dotted line shows the results without rinsing. There was no evidence of a downward trend or of a significant difference between both conditions. Therefore no adaptation has accumulated through the stimuli of a sequence (Kroeze, 1982c).

intervals just long enough to prevent adaptation. In this way a lot of exposure can be accumulated over a short time period. Immediately after such a sequence of identical, habituating stimuli, we may offer a binary mixture of which one component is the habituating stimulus. If suppression release occurs after such exposure it cannot be ascribed to 'adapting away' one of the components, since we carefully prevented adaptation throughout the series of stimuli. Such experiments were in fact carried out (Kroeze, 1982b, 1983). A typical result is shown in Fig. 3.

Suppression-release occurs in Bartoshuk's (1975) and Lawless' (1979) suppression release experiments. However, in these cases we may draw a more extreme conclusion concerning the location of the mixture suppression mechanism. These experiments exclude the peripheral nature of mixture suppression. They suggest that mixture suppression occurs at a level central to the habituation locus: it is a brain mechanism not a receptor mechanism. With brain mechanisms we may think of distinct neural systems that have inhibitory projections to each other. However, these systems are not likely to project directly to each other. Direct mutual inhibition would not enable us to distinguish between adaptation and habituation as determinants of suppression release.

Bilateral stimulation

There is another way to obtain knowledge of the locus of mixture suppression. This is by bilateral stimulation of the tongue. With this method we tried to separate a peripheral and a central contribution to mixture interaction (Kroeze, 1977; Kroeze & Bartoshuk, 1985). The

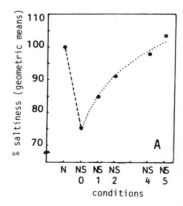

FIG. 3. Suppression release of saltiness after repeated sweet stimuli. The difference between N and NS-0 represents saltiness suppression by sucrose. 1, 2, 4 and 5 indicate the number of successive sucrose stimuli preceding the mixture.

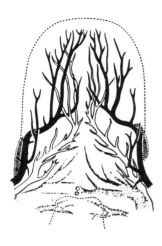

FIG. 4. Neural innervation of the embryonic tongue. The nerve supply is strictly ipsilateral and does not cross the medial plane (Vij & Kanagasuntheram, 1972).

tongue was divided into two halves, left and right. This was accomplished by a tongue chamber with a central bar pressing on the tongue's mid line. Then the tongue was stimulated with taste solutions on the left or right side or on both sides simultaneously. The tongue itself offers an appropriate opportunity for such an approach. The neural and vascular innervation of the left and right side of the tongue are ipsilateral as is illustrated in Fig. 4.

The first mid-line crossings of neural information occur at levels higher than the nucleus of the solitary tract, probably not earlier than at the thalamic level (Gerebtzoff, 1939; Norgren & Leonard, 1973). Comparison of conditions in which stimulus components are *mixed* on the tongue (thus allowing for interaction at the receptor site) with conditions where components are spatially *separated* by the tongue's midline may provide us with a partition ratio of central to peripheral suppression. If mixture suppression is entirely central, no difference between conditions where components are mixed versus separated would be expected. On the other hand, if mixture suppression is entirely peripheral, the split-tongue conditions cannot lead to suppression. In this experiment two mixtures were investigated: 0·001 mol/litre quinine hydrochloride (QHCl)/0·32 mol/litre sucrose and 0·001 mol/litre QHCl/0·32 mol/litre NaCl (Kroeze & Bartoshuk, 1985). When QHCl and sucrose were mixed on the same tongue half, bitterness was suppressed by 37%. With sucrose on the opposite tongue half, bitterness was suppressed by 31%. Although mixture suppression was statistically significant in both conditions, the *difference* was not statistically significant. Thus the suppression of bitterness was not significantly increased by adding the opportunity for peripheral suppression to the opportunity for central suppression. This suggests that QHCl-bitterness is centrally suppressed by sucrose. For the QHCl/NaCl mixture the results were different. The 'same tongue half' condition produced 69% bitterness suppression and the split-tongue condition 23%. Both percentages indicated statistically significant amounts of mixture suppression. But this time, the difference between these percentages was also significant. Thus it was concluded that bitterness suppression in the case of a QHCl/NaCl mixture resulted from a combination of 46% peripheral and 23% central bitterness suppression. This conclusion assumes that the combination rule of both suppression contributions is simple addition. Combination of qualitatively different taste stimuli does not always result in mixture suppression when the stimuli are of near-threshold intensity. This is shown by the following experiment (Kroeze, 1977). First the individual threshold concentrations of sucrose and NaCl were determined for each member of a group of ten subjects. Then sucrose and NaCl of the appropriate threshold concentration per subject were mixed in the following proportions: NaCl/sucrose: 100/0, 80/20, 60/40, 50/50, 40/60, 20/80 and 0/100. Stimulation with these near-threshold mixtures, in a split-tongue set up, resulted in a lower absolute threshold than might have been expected on the basis of the unmixed components, but this was true only in the 'mixed on the tongue' condition and not in the split-tongue condition (Kroeze, 1977). So sensitivity enhancement with these low-intensity mixtures was entirely dependent on peripheral mixing of components. It also appeared that the enhancement was strongest when the threshold concentrations of NaCl and sucrose were of about equal proportion in the mixture. Although it is not known how this enhancement is caused, the following explanation can be offered. Pfaffmann (1955) showed that taste nerves have multiple sensitivity. This means that one and the same nerve fibre may be activated by several qualitatively different stimuli applied to taste cells. However, taste cells activated by the two substances in the split-tongue condition do not spatially summate into common fibres. Thus, at threshold, relative spatial density is more important than the absolute amount of activity in the taste system, and if density is replaced by amount, then detection declines. This does not logically prove that the increased sensitivity is peripherally determined, but it does suggest that, with low concentrations, spatially separated adsorption of stimulus molecules is less efficient in terms of neural activity than spatially mixed adsorption. If the distances between fibres on the tongue are replicated in the brain taste area, then peripheral proximity of NaCl and sucrose may be a necessary condition and the actual summation may be carried out at either level. In order to understand fully the meaning of these threshold results, we must know how central taste cells combine their activity. The experiments show that not all mixture suppression is central. Substances may compete for common receptor sites as many sugars probably do (Shallenberger & Acree, 1971; Kier, 1972). This competitive adsorption may show itself in electrophysiological recordings as suppression. Beidler (1971, 1978) and Heck and Erickson (1973) modelled this type of suppression mathematically and these competitive mixture models predict receptor potentials and first-order neural activity. However competitive binding is not reflected directly in sensation, particularly not when central suppression components are added at a higher processing level. Competitive binding theories are receptor theories and are therefore not suited to predict behaviour. Nevertheless some authors have successfully predicted sensation by Beidler's mixture model (Curtis et al., 1984). But in these cases the mixture components were of similar quality. With similar qualities competitive suppression prevails and accordingly mixture models predict a compromise of the mixture components (Frijters & Oude Ophuis, 1983), i.e. perceived mixture intensity lies between the perceived intensities of the components (De Graaf & Frijters, 1987). When the compression or acceleration of the psychophysical function is dependent on receptor dynamics, the combined receptor processes as well as the combined psychophysical functions predict sensation. Peripheral mixture theories may be true for similarly tasting substances, but they are inappropriate for taste substances of different quality.

Competition equations make very precise predictions about the response if and only if the parameters of the formula can be specified. There is no problem with the stimulus concentrations and the binding constants of the mixture components. However, CR_{max}, the stimulus concentration that evokes the largest possible reported sensation, is much more difficult to obtain. A minor shift in decision criterion as to what is considered the maximum value on the response axis may lead to a large difference on the concentration axis of the psychophysical function. De Graaf and Frijters (1986) tried to bypass this problem by algebraic elimination of CR_{max} and the binding constants. This resulted in a weak test of the competition equation, since only relative judgments of stimulus pairs were allowed under the impoverished theoretical restrictions. Maes (1985) used a different procedure to overcome uncertainty with respect to CR_{max}. Starting from the assumption that Beidler's taste equation is correct ($R/R_{max} = 1/(1+K/C)$, in which R = response magnitude, K = binding constant and C = concentration of stimulus) he argued that the point of maximum slope in the sigmoid semilog plot is located halfway to the maximum response. Since the sigmoid function is relatively linear around this inflexion point, a linear plot through two points in this region would give us the R_{max} value by extrapolation. Maes recommends this procedure to improve best stimulus classification of taste neurons, i.e. to use it in a relative way. As we have noticed, De Graaf and Frijters' (1986) solution was also a relative solution. McBride (1987) demonstrated the *relative* use of Maes' approach in a Beidler plot of the sweetnesses of three sugars. From an absolute point of view the problem is not solved in either of these ways, since R_{max} cannot be obtained independently from R. The problem remains that responses cannot be predicted in an absolute way with an equation that contains one of the responses to be predicted (R_{max}) as an unknown parameter. As for psychophysical responses there seems to be only one reasonable advice: don't predict sensation from receptor activity alone. It ignores the brain and does not consider the response-output function. A sequential model is needed, which chains the successive contributions from ever higher levels into a compound predictor of response intensity. The weight constants of each contribution and the combination rules have to be determined in order to make such a model work. De Graaf and Frijters (1988) made an attempt to outline such a model.

INFORMATION PROCESSING IN TASTE

Compared to vision, taste information-processing capacity is modest. When subjects are asked to rank order a large number of taste stimuli with respect to intensity, it may be observed that they spontaneously break down the total number into smaller groups.

Miller (1956) estimated the information processing capacity of several sensory systems. He calculated that, in taste, the capacity of the channel equalled 1·8 bit, which compares to a mean of 3·5 alternatives. Thus complex judgments involving more than three alternatives may begin to pose problems. For the purpose of sensory evaluation, two- and three-stimulus procedures generally remain within the limits of human ability.

Exceeding the information processing capacity creates considerable between-subject variance since at least two additional factors come into play: individual information capacity differences and differences in the way subjects break down a series of stimuli into appropriate subgroups. Furthermore, instead of being controlled by the experimenter, stimulus duration (tasting longer) and stimulus frequency (tasting more often) are progressively controlled by the subjects in their effort to produce the required response. In some cases information overload may be deliberately built into the procedure because the investigator wants to mimic as closely as possible the sensory, decisional and choice factors occurring in real-life consumer behaviour. The fact that the mean judged number of different tastes in a mixture of sucrose, quinine, salt and citric acid is less than four may be due to the limited capacity of the working memory for taste. Figure 5 shows that with increasing physical complexity a taste mixture's perceived complexity decreases. The mean complexity of an unmixed substance is systematically higher than 1, most likely because of side tastes. With two components the number of reported tastes equals 2 and thus apparently no information is lost. But with more than two components, reported complexity shrinks.

The fact that subjects report less than four components when tasting

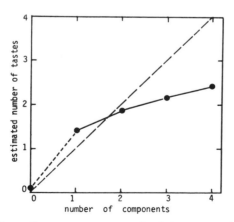

FIG. 5. The number of components perceived in a complex taste mixture increases at a lower rate than the actual number.

a mixture of four tastes may also be interpreted as an indication of partial synthesis in taste. In vision the sensation of separate colours in a mixture is lost and instead a new colour is seen. In other words, if four different wavelengths, each associated with a unique unidimensional colour sensation, are put into a mixture the reported number of colours is lower than four — typically it is only one. It is said that there is complete synthesis in vision. On the synthetic-analytic dimension vision is very much at the synthetic side. Audition, on the other hand, is very much analytic. According to Erickson (1982) taste is in between, i.e. partially synthetic and partially analytic. This would imply an inverse relationship between the perceived number of tastes in a complex mixture and the amount of mixture suppression. An experiment to test this showed that mixture suppression is not related at all to perceived complexity (Kroeze, 1984). Instead of discussing the loss of reported complexity in terms of synthesis and analysis, it may perhaps be more fruitful to interpret it in terms of a limited ability to process and report taste information.

Onset of stimulus components in complex mixtures
Stimulus components with strong spatio-temporal relationships may be involved in two processes: mutual masking (Breitmeyer & Ganz, 1976) or patterning (Koffka, 1935). From most senses, examples of simple stimuli forming a pattern can easily be found. For example, separate tones sufficiently close in time form a melodic pattern, combined odours constitute perfumes, visual stimuli organize into pictures. On the other hand, masking also occurs abundantly in perception: a voice cannot be heard because of traffic noise (auditory intensity masking) or because of other voices heard simultaneously (auditory frequency masking). Very well known are visual backward masking phenomena: a briefly displayed (5 ms) visual stimulus (e.g. a number) followed within 200 ms or sooner by a stimulus at the same location is completely or partially masked, i.e. the subject does not see the number.

With attentional effort subjects may succeed in analysing complex stimuli and counteract masking and patterning to a certain degree.

Kuznicki and Turner (1988) showed that mutual masking of taste stimuli in a mixture is counteracted by differences in taste onset times of the components. Recognition of 1 ml of either a bitter taste (0·1M caffeine) or a salty taste (0·3M NaCl) in the mixture of both stimuli was related to the reaction time required to recognize the unmixed components. It turned out that caffeine bitterness required a much longer recognition time than NaCl saltiness. What was more interesting was the fact that the salty target, when mixed with the bitter stimulus, was detected far more often than the bitter target when mixed with the salty stimulus. When the subjects were forced to respond within a deadline of 1 s the percentage of correct bitterness recognitions in the mixture was only 50% as compared to 88% in the unmixed condition. This contrasted to the percentages of correct saltiness recognitions which did not differ between the mixed (93%) and unmixed (95%) condition. When the subjects were allowed later response deadlines (up to 2·5 s) the difference in bitterness recognition between mixed (77%) and unmixed (90%) stimuli decreased significantly. These experiments of Kuznicki and Turner show the following important points relevant to our discussion: first, quality detection is masked in mixtures; secondly, sensation onset differences are a discriminative cue for taste quality in mixtures; thirdly, subjects can undo recognition masking by attentional shifts if they are given extra response time.

SUMMARY

Taste as a complex sensory modality only partially contributes to the perceptual patterns evoked by foods and beverages. Perceptual patterns, such as those elicited by taste substances are the conscious result of largely unconscious and quick perceptual integration. Taste patterns may be analysed by attentional effort into a set of perceptual components, which do not necessarily coincide with the physical composition of the taste stimulus.

Taste adaptation is a peripheral mechanism that adjusts the differential sensitivity to the prevailing intensity level. In normal eating, taste adaptation is incomplete and reflects the equilibrium between adaptation and recovery. Mixture suppression has a peripheral competitive component and a central perceptual-masking component. The effectiveness of analysing mixtures depends on the effort invested, the degree of patterning, and cues such as temporal differences in stimulus-onset of components.

REFERENCES

Abrahams, H., Krakauer, D. & Dallenbach, K. M. (1937). Gustatory adaptation to salt. *Am. J. Psychol.,* **49,** 462-9.

Anderson, C. D. (1955). The effect of subliminal salt solutions on taste thresholds. *J. Comp. Physiol. Psychol.*, **48**, 164–6.

Bartoshuk, L. M. (1975). Taste mixtures: is mixture suppression related to compression? *Physiol. Behav.*, **14**, 643–9.

Bartoshuk, L. M. (1978). Sweet taste of dilute NaCl: psychophysical evidence for a sweet stimulus. *Physiol. Behav.*, **21**, 609–13.

Bartoshuk, L. M., McBurney, D. H. & Pfaffmann, C. (1964). Taste of sodium chloride after adaptation to sodium chloride: implications for the 'water taste'. *Science*, **143**, 967–8.

BeebeCenter, J. G., Rogers, M. S., Atkinson, W. H. & O'Connell, D. N. (1959). Sweetness and saltiness of sucrose and NaCl as a function of solutes. *J. Exp. Psychol.*, **57**, 231–4.

Beidler, L. M. (1971). Taste receptor stimulation with salts and acids. In *Handbook of Sensory Physiology*, Vol. IV, part 2, ed. L. M. Beidler. Berlin, New York, Springer. pp. 200–20.

Beidler, L. M. (1978). Biophysics and chemistry of taste. In *Handbook of Perception*, Vol. VIA, ed. E. C., Carterette & M. P. Friedman. Academic Press, New York, pp. 21–49.

Breitmeyer, B. G. & Ganz, L. (1976). Implications of sustained and transient channels for theories of visual pattern masking, saccadic suppression, and information processing. *Psychol. Rev.*, **83**, 1–36.

Burdach, K. J. & Doty, R. L. (1987). The effect of mouth movements, swallowing, and spitting on retronasal odor perception. *Physiol. Behav.*, **41**, 353–6.

Cardello, A. V. & Murphy, C., Magnitude estimates of gustatory quality change as a function of sodium concentrates of simple salts. *Chem. Sens. Flav.*, **2**, 327–39.

Coombs, C. H. (1979). Models and methods for the study of chemoreception hedonics. In *Preference Behaviour and Chemoreception*, ed. J. H. A. Kroeze. IRL-Press, London. pp. 149–70.

Curtis, D. W., Stevens, D. A. & Lawless, H. T. (1984). Perceived intensity of the taste of sugar mixtures and acid mixtures. *Chem. Sens.*, **9**, 107–20.

De Graaf, C. & Frijters, J. E. R. (1986). A psychophysical investigation of Beidler's mixture equation. *Chem. Sens.*, **11**, 295–314.

De Graaf, C. & Frijters, J. E. R. (1987). Sweetness intensity of a binary sugar mixture lies between intensities of its components, when each is tasted alone and at the same total molarity as the mixture. *Chem. Sens.*, **12**, 113–29.

De Graaf, C. & Frijters, J. E. R. (1988). Interrelationships among sweetness, saltiness and total intensity of sucrose, NaCl and sucrose/NaCl mixtures. In *Psychophysical Studies of Mixtures of Tastants*. Thesis, C. de Graaf, Agricultural University Wageningen, Wageningen. pp. 158–94.

Diamant, H., Oakley, B., Ström, L. & Zotterman, Y. (1965). A comparison of neural and psychophysical responses to taste stimuli in man. *Acta Physiol. Scand.*, **64**, 67–74.

Doty, R. L. & Kimmelman, C. P. (1986). Smell and taste and their disorders. In *Diseases of the Nervous System*, ed. A. K. Ashbury, G. M. McKhann & W. I. McDonald. Saunders, Philadelphia. pp. 466–78.

Erickson, R. P. (1978) The role of 'primaries' in taste research. In *Olfaction and*

Taste VI, ed. J. Le Magnen and P. MacLeod. IRL Press, Washington D.C. and London. pp. 369-76.
Erickson, R. P. (1982). Studies in the perception of taste: do primaries exist? *Physiol. Behav.,* **28,** 57-62.
Erickson, R. P. & Covey, E. (1980). On the singularity of taste sensations: what is a taste primary? *Physiol. Behav.,* **25,** 527-33.
Fabian, F. W. & Blum, H. B., Relative taste potency for some basic food constituents and their competitive and compensatory action. *Food Res.,* **8,** 179-93.
Frijters, J. E. R. & Oude Ophuis, P. A. M. (1983). The construction and prediction of psychophysical power functions for the sweetness of equiratio sugar mixtures. *Perception,* **12,** 753-67.
Ganzevles, P. G. J. & Kroeze, J. H. A. (1987a). Cross adaptation in taste measured with a filter-paper method. *Chem. Sens.,* **12,** 341-53.
Ganzevles, P. G. J. & Kroeze, J. H. A. (1987b). The sour taste of acids. The hydrogen ion and the undissociated acid as sour agents. *Chem. Sens.,* **12,** 563-76.
Gent, J. & McBurney, D. H. (1978). Time course of gustatory adaptation. *Percept. Psychophys.,* **23,** 171-5.
Gerebtzoff, M. A. (1939). Les voies centrales de la sensibilité et du gout et leurs terminaisons thalamiques. *Cellule,* **48,** 91-146.
Gillan, D. J. (1982). Mixture suppression: the effect of spatial separation between sucrose and NaCl. *Percept. Psychophys.,* **32,** 504-10.
Hahn, H. (1934). Die Adaptation des Geschmacksinnes. *Z. Sinnesphysiol.,* **65,** 105-45.
Heck, G. L. & Erickson, R. P. (1973). A rate theory of gustatory stimulation, *Behav. Biol.,* **8,** 687-712.
Helson, H. (1964). *Adaptation-level theory.* New York, Harper & Row.
Hyman, A. (1983). The influence of color on taste perception of carbonated water preparations. *Bull. Psychon. Soc.,* **21,** 145-8.
Indow, T. (1969). An application of the τ-scale of taste: interactions among the four qualities of taste. *Percept. Psychophys.,* **4,** 347-51.
Johnson, J. L. & Clydesdale, F. M., Perceived sweetness and redness in colored sucrose solutions. *J. Food Sci.,* **47,** 747-52.
Johnson, J. L., Dzendolet, E., Damon, R., Sawyer, M. & Clydesdale, F. M. (1982). Psychophysical relationships between perceived sweetness and color in cherry-flavored beverages. *J. Food Prot.,* **45,** 601-6.
Kahn, S. G. (1951). Taste perception. Individual reactions to different substances. *Trans. Illinois State Acad. Sci.,* **44,** 263-9.
Kahneman, D. (1973). *Attention and Effort.* Prentice Hall, Englewood Cliffs, N.J.
Kamen, J. M., Pilgrim, F. J., Gutman, N. J. & Kroll, B. J. (1961). Interactions of suprathreshold stimuli. *J. Exp. Psychol.,* **62,** 348-56.
Keidel, W. D., Keidel, U. O., & Wigand, M. E. (1961). Adaptation: loss or gain of sensory information? In *Sensory Communication,* ed. W. A. Rosenblith. John Wiley, New York. pp. 319-38.
Kier, L. B. (1972). A molecular theory of sweet taste. *J. Pharmac. Sci.,* **61,** 1394-6.
Koffka, K. (1935). *Principles of Gestalt Psychology.* Harcourt, New York, p. 35.

Köhler, W. (1947). *Gestalt Psychology* (revised edition). Liveright, New York.
Kroeze, J. H. A. (1977). Taste thresholds for bilaterally and unilaterally presented mixtures of sugar and salt. In *Olfaction and Taste VI*, ed. J. Le Magnen & P. MacLeod. IRL-Press, London, p. 486.
Kroeze, J. H. A. (1978). The taste of sodium chloride: masking and adaptation. *Chem. Sens. Flav.*, **3**, 443-9.
Kroeze, J. H. A. (1979). Masking and adaptation of sugar sweetness intensity. *Physiol. Behav.*, **22**, 347-51.
Kroeze, J. H. A. (1980). Masking in two and three-component taste mixtures. In *Olfaction and Taste VII*, ed. H. van der Starre. IRL-Press, London, p. 435.
Kroeze, J. H. A. (1982a). The relationship between the side taste of masking stimuli and masking in binary mixtures. *Chem. Sens.*, **7**, 23-37.
Kroeze, J. H. A. (1982b). After repetitious sucrose stimulation saltiness suppression in NaCl-sucrose mixtures is diminished: implications for a central mixture-suppression mechanism. *Chem. Sens.*, **7**, 81-92.
Kroeze, J. H. A. (1982c). Mixture suppression depends on a central inhibitory mechanism. In *Determination of Behaviour by Chemical Stimuli*, ed. J. Steiner & J. Ganchrow. IRL-Press, London. pp. 161-74.
Kroeze, J. H. A. (1982d). The influence of relative frequencies of pure and mixed stimuli on mixture suppression in taste. *Percept. Psychophys.*, **31**, 276-8.
Kroeze, J. H. A. (1983). Successive contrast cannot explain suppression release after repetitious exposure to one of the components of a taste mixture, *Chem. Sens.*, **8**, 211-23.
Kroeze, J. H. A. (1984). The subjective complexity of taste stimuli. In *Abstracts of ECRO VI*. ECRO, Lyon, p. 74.
Kroeze, J. H. A. & Bartoshuk, L. M. (1985). Bitterness suppression as revealed by split-tongue taste stimulation in humans. *Physiol. Behav.* **35**, 779-83.
Kuznicki, J. T. & Ashbaugh, N. (1979). Taste quality differences within the sweet and salty taste categories. *Sens. Proc.*, **3**, 157-82.
Kuznicki, J. T. & Turner, L. S. (1988). Temporal dissociation of taste mixture components. *Chem. Sens.*, **13**, 45-62.
Lawless, H. T. (1979). Evidence for neural inhibition in bitter-sweet taste mixtures. *J. Comp. Physiol. Psychol.*, **93**, 538-47.
Lawless, H. T. (1982a). Adapting efficiency of salt-sucrose mixtures. *Percept. Psychophys.*, **32**, 419-22.
Lawless, H. T. (1982b). Paradoxical adaptation in taste mixtures. *Physiol. Behav.*, **25**, 142-52.
Loeb, J. (1901). The poisonous character of pure NaCl solution. *Am. J. Psychol.*, **3**, 327.
Maes, F. W. (1985). Improved best-stimulus classification of taste neurons. *Chem. Sens.*, **10**, 35-44.
McBride, R. L. (1987). Taste psychophysics and the Beidler equation. *Chem. Sens.*, **12**, 323-32.
McBurney, D. H. (1984). Taste and olfaction: sensory discrimination. In *Handbook of Physiology*, Vol. III: *The Nervous System*, ed. I. Darian-Smith, part 2: *Sensory Processes*. American Physiological Society, Bethesda, Maryland, pp. 1067-86.

McBurney, D. H. (1986). Taste, smell, and flavor terminology: taking the confusion out of fusion. In *Clinical Measurement of Taste and Smell*, ed. H. L. Meiselman & R. S. Rivlin. MacMillan, New York, pp. 117-25.
McBurney, D. H., Kasschau, R. A. & Bogart, L. M. (1967). The effect of adaptation on taste jnd's. *Percept. Psychophys.*, **2**, 175-8.
Miller, G. A. (1956). The magical number seven, plus or minus two: some limits on our capacity for processing information. *Psychol. Rev.*, **63**, 81-97.
Moskowitz, H. R. (1972). Perceptual changes in taste mixtures. *Percept. Psychophys.*, **11**, 257-62.
Mozell, M. M., Smith, B. P., Smith P. E., Sullivan Jr, R. J. & Swender, P. (1969). Nasal chemoreception in flavor identification. *Arch. Otolaryngol.*, **90**, 131-7.
Norgren, R,. & Leonard, C. M. (1973). Ascending central gustatory pathways. *J. Comp. Neurol.*, **150**, 217-38.
O'Mahony, M. (1972*a*). The interstimulus interval for taste: 1. The efficiency of expectoration and mouthrinsing in clearing the mouth of salt residuals. *Perception*, **1**, 209-15.
O'Mahony, M. (1972*b*). The interstimulus interval for taste: 2. Salt taste sensitivity drifts and the effects on intensity scaling and threshold measurement. *Perception*, **1**, 217-22.
O'Mahony, M., Kingsley, L., Harji, A. & Davies, M. (1976). What sensation signals the salty taste threshold? *Chem. Sens. Flav.*, **2**, 177-88.
Pangborn, R. M. (1960). Taste interrelationships. *Food Res.*, **25**, 245-56.
Pangborn, R. M. (1961). Taste interrelationships II: Suprathreshold solutions of sucrose and citric acid. *J. Food Sci.*, **26**, 648-55.
Pangborn, R. M. (1962). Taste interrelationships III: Suprathreshold solutions of sucrose and sodium chloride. *J. Food Sci.*, **27**, 495-500.
Pangborn, R. M. (1967). Some aspects of chemoreception in human nutrition. In *The Chemical Senses and Nutrition*, ed. M. R. Kare & O. Maller. John Hopkins Press, Baltimore, Md, pp. 45-60.
Pangborn, R. M. & Trabue, I. M. (1967). Detection and apparent taste intensity of salt-acid mixtures in two media. *Percept. Psychophys.*, **2**, 503-9.
Pfaffmann, C. (1955). Gustatory nerve impulses in rat, cat and rabbit. *J. Neurophysiol. (Bethesda)*, **18**, 429-40.
Polich, J. (1987). Comparison of P300 from a passive tone sequence paradigm and an active discrimination task. *Psychophysiology*, **24**, 41-6.
Renqvist, Y. (1919). Über den Geschmack. *Skand. Arch. Physiol.*, **38**, 97-201.
Richter, C. P. & McLean, A. (1939). Salt thresholds in humans. *Am. J. Physiol.*, **126**, 1-6.
Sams, M., Alho, K. & Näätänen, R. (1984). Short-term habituation and dishabituation of the mismatch negativity of the ERP. *Psychophysiology*, **21**, 434-41.
Schechter, P. J. & Henkin, R. I. (1974). Abnormalities of taste and smell after head trauma. *J. Neurol. Neurosurg. Psychiat.*, **37**, 802-10.
Shallenberger, R. S. & Acree, T. E. (1971). Chemical structure of compounds and their sweet and bitter taste. In *Handbook of Sensory Physiology*, Vol. IV, part 2. ed. L. M. Beidler. Springer, New York. pp. 221-77.
Stone, H., Oliver, S. & Kloehn, J. (1969). Temperature and pH-effects on the

relative sweetness of suprathreshold mixtures of dextrose fructose. *Percept. Psychophys.*, **5**, 257–60.

Vij, S. & Kanagasuntheram, R. (1972). Development of the nerve supply to the human tongue. *Acta Anat.*, **81**, 466–77.

Von Békésy, G. (1965). The effect of adaptation on the taste threshold observed with a semi-automatic gustometer. *J. Gen. Physiol.*, **48**, 481–8.

Zotterman, Y. (1971). The recording of the electrical response from human taste nerves. In *Handbook of Sensory Physiology,* Vol. IV, Part 2, ed. L. M. Beidler. Springer, New York. pp. 102–15.

4

Applications of Experimental Psychology in Sensory Evaluation

HARRY LAWLESS

Department of Food Science, New York State College of Agriculture and Life Sciences, Cornell University, Ithaca, New York 14853, USA

INTRODUCTION

In this chapter some of the relationships between the practice of sensory evaluation and selected topics within experimental psychology that relate to methods are discussed. In doing so, the author has drawn on his experiences as a product evaluation practitioner serving on the front lines of product development. In order to understand the relationship between sensory evaluation practice and psychological theory, it helps to understand how sensory methods are applied in the commercialization of real products under cost constraints and tight deadlines. This chapter may serve to bring to the attention of industrial sensory practitioners some *opportunities* for greater application of psychophysical techniques and principles of human perception. The application of these principles may serve to advance the field of industrial sensory practice.

There are three fundamental questions asked about products in sensory evaluation. The first concerns whether two products are perceivably different. This question in its simplest form is attacked by means of discrimination tests like the duo-trio and triangle procedures. The assessment of discriminability bears a logical resemblance to the traditional psychophysical questions about just noticeable differences. Furthermore, since the absolute threshold is a special case of a difference threshold, practical detectability questions such as the detection of off-notes in the practice of quality control fall into this category as well.

The second major question in sensory evaluation concerns the acceptability of a product. Two products may be discriminably different, but equally acceptable. The hedonic characteristics of a product are closely related to its success in the marketplace, and thus receive a great deal of attention from product evaluation specialists.

The third type of question concerns the analytical specification of the sensory characteristics of a product. What sensory qualities are present, and how intense are they? These questions, which quantify what something tastes or smells like, are most often addressed by a class of techniques called descriptive analysis, in which sensory attributes are rated on scales. Two parts of this process involve psychological issues. First, there are linguistic-conceptual questions about how to identify and define the sensory characteristics which are present and which are likely to differentiate products. Practitioners of descriptive analysis need to consider the linguistic and phenomenological questions about how best to talk about experiences and perceived qualities of sensations. Second, there are questions concerning measurement techniques, such as the choice of scaling method, calibration of panelists, possible contextual biases, and so on.

This chapter discusses five areas of psychological research which have an important influence upon the conduct of sensory evaluation. These sections will include: (1) the discrimination of differences, (2) the scaling of sensory intensity, (3) the identification of sensory attributes, (4) perceptual and memory phenomena involved in product description, and (5) the assessment of methodologies *per se*. This last area is important insofar as the field of psychology has determined criteria for 'what is a good assessment device', yet these criteria have not generally been applied in evaluation of methodologies for industrial practice.

DIFFERENCE TESTING, DIFFERENCE THRESHOLDS AND SCALING DIFFERENCES

The simple question of whether two products are perceivably different is addressed in sensory evaluation for a variety of reasons. One common question concerns the stability or shelf-life of a product. A product which has been aged for a certain period is compared to some control, often a freshly-made version of the item or one stored under optimal conditions. Reformulation or cost-reduction efforts will often call for a simple difference test. For example, it may be necessary to determine whether an ingredient substitution produces any change in the sensory characteristics of a product. Some form of difference testing is routinely needed in quality control of manufacturing, especially when there is variability in processing conditions or in formulation. The detection of off-flavors may be viewed as a special case of a difference test, one which has been reduced to the threshold situation. Historically, difference tests have been used as a panelist screening tool, i.e. to identify individuals with normal or exceptional acuity in order to qualify them for quality control panels or other sensory analysis duties (Helm & Trolle, 1946; Bressan & Behling, 1977). The common theme in these tests is that they are all variations of forced-choice tasks in which the observer is asked to choose the different (or same) item from a number of alternatives (e.g. triangle and duo-trio tests).

Forced-choice procedures are also used in the assessment of just-noticeable differences (JNDs) or difference thresholds in the psychophysical laboratory. The sensory difference test is logically related to JNDs in that at least one sensory characteristic in the test product must be at least one JND different from the control sample in order for a difference to be detected. There are major differences in practice, however. The sensory difference test is usually a go/no-go procedure, used for making decisions about a single pair of products. In contrast, psychophysical assessment of JNDs specifies the evidence for a detectable difference at several points along a physical continuum. More than two stimuli are assessed. The question is not framed in terms of difference or equivalence, but as how differences become more perceivable as the test stimulus moves farther from the control stimulus. The difference is said to be detected when some arbitrary convention is surpassed, such as correct choices on 50% or 75% of trials. The sensory difference test, on the other hand, reports a difference relative to chance performance in the forced-choice procedure at a single point. When the observed proportion of correct choices exceeds the proportion expected by chance to such a degree that the observed proportion would occur less than 5% of the time assuming that no difference really exists, then the products are deemed different. Conversely, when the observed proportion fails to exceed the chance level (by an improbably small amount), the products are often deemed to be perceptually equal. This is a curious conclusion based upon a failure to reject the null hypothesis, given that the risk for Type II error is rarely estimated in these situations.

The notion of increasing evidence for a difference as stimuli move farther apart is embraced by the theory of signal detection. In this theory, the notion of a threshold is abandoned and perceivability of a difference is viewed as a continuous function of the physical difference between stimuli. This can be a difficult concept for people who wish to make a decision about whether products are different or the same. It replaces the statistical decision process with a measure of difference based on the proportion of correct choices, rather than on the degree of confidence that a Type I error has been avoided. The relationships of signal detection measures of degree of difference and the triangle test have been thoroughly explored in the writings of Frijters (1979a, b).

This work has two implications. First, difference testing should not be viewed in a go/no-go fashion, in spite of the need to make actionable recommendations to clients based on the outcome of a difference test. Instead, there is a graded degree of difference or at least a graded degree of evidence for a difference. Second, the industrial practitioner should consider the signal detection measures as a means of quantifying and reporting the degree of difference observed. The notion in the applied literature that a degree of difference cannot be quantified from the triangle test is patently false; the signal detection measures of difference are clearly related to the outcome of forced-choice tests in both theory and in practice, with tables available for mathematical conversion (Frijters, 1979b).

However, product differences which are derived from forced-choice tests may not necessarily duplicate the pattern found with methods such as the direct scaling of single attributes (Lawless & Schlegel, 1984). Real products are multidimensional in character, unlike the simple unidimensional stimuli routinely used in the psychophysical laboratory. In applying the signal detection measures to industrial difference tests it is tempting to assume that the observer has been paying attention to the one underlying dimension or characteristic which is different for the two products. However, different observers may decide to pay attention to different sensory attributes of multidimensional stimuli, and may miss the difference entirely. This occurs not because the difference is undetectable to the observer, but because the focus of attention is elsewhere. Thus observed performance may underestimate the true degree of difference that is present. The signal detection analysis of difference tests has been extended to models which take into account the multidimensional nature of products (Ennis & Mullen, 1985).

THE IMPORTANCE OF SCALING

Scaling is a powerful and frequently employed tool in sensory evaluation. The nature of scaling procedures — their validity and the types of psychophysical functions they produce — are issues that have received intense debate in the psychological literature. In spite of this discussion, from a practical perspective, category scales, magnitude estimation and line scales are all about equally sensitive when it comes to finding product differences (e.g. Lawless & Malone, 1986). This assumes that the scales are used in a reasonable fashion, e.g. with sufficient scale points for fine discriminations and with some awareness of biases and context effects on the part of the experimenter.

Why be concerned with the relationship of the perceived intensity of the sensory characteristics of a product to its physical characteristics? If people like a product, it is reasonable to assume that the product has pleasing sensory characteristics in acceptable combinations and at appropriate intensity levels. In other words, an (unspoken) model forms the foundation of much of our thinking about the acceptability of a product. This model includes terms related to the perceived intensity of various critical sensory characteristics. One such simple model is a weighted polynomial (Moskowitz, 1983), in which the acceptability of a product, A, is predicted by an equation of the form

$$A = k_{i1}S_1 + k_{j2}S_2 + \ldots$$

where S is the sensory intensity of an attribute and k_{i1} is a constant. This equation may include higher order terms such as

$$\ldots + k_{i2}S_1^n \ldots$$

where a negative k_{i2} term and $n = 2$ will produce the inverted U-shaped curve typical of hedonic functions with a single optimum or 'bliss point'. The model may also include multiplicative interaction terms such as

$$k_{i3}S_1^{n_1} * k_{j3}S_2^{n_2} \ldots$$

In the case where $n_1 = 1$ and $n_2 = -1$, we have a ratio of the two sensations, a common characteristic of fruit-derived foods and beverages such as wine, in which sweet-to-sour ratios are more important than absolute sweetness or sourness (Moskowitz, 1983).

The sensation levels, S_i, can be related to the physical concentration of ingredients by power functions or some alternative dose–response function such as the Beidler equation (McBride, 1987). Thus it should be possible, in principle, to predict hedonic optima on the basis of ingredients. In practice, this can sometimes be achieved in simple systems using response surface experiments or through application of multiple regression techniques. Thus consumer-defined acceptability should be predictable on the basis of sensory characteristics that are analytically measured through techniques such as descriptive analysis. In turn, those sensory characteristics can be related to ingredient levels which can be systematically manipulated by a development chemist to produce an optimal product. Scaling is the tool needed to achieve this. There are, however, some pitfalls.

In order for such a regression-type polynomial model to be useful, certain measurement requirements must be met or at least approximated. The terms A and S_i must be measured with a reasonable degree of accuracy and precision. A primary reason for the importance of accurate scaling techniques in sensory evaluation is that the accuracy and precision of these measurements sets an upper limit on the amount of variance explainable by any model relating consumer acceptability to sensory attributes. A corollary to this requirement is that there should be relatively good inter-subject agreement. In the case of hedonic judgment, this is never guaranteed. Individual differences are the rule. Of course, one always has the option of adding an additional layer of complexity and developing specific models for segmentable groups of respondents.

Additional requirements are that the sensory characteristics and their interactive mechanisms are fully specified and quantified in the measurement procedure. This has two parts. First, no important determining factor can be left out of the measurement of sensory attributes. In regression terms, all those attributes which weight heavily onto overall acceptability must be part of the model. This is one reason for careful descriptive analysis procedures. Second, interactions among sensory attributes should also be determined. Such interactions include phenomena like mixture suppression, synergy between ingredients, as well as higher order perceptual interactions that tend to cause sensory characteristics to be perceived as a pattern or unit, as in the case of the sweet/sour ratio that is important in wines. Few gestalt-like interactions have been identified for the chemical senses, although many probably wait to be discovered.

There are other advantages to good scaling in sensory evaluation of products. If an ingredient can be systematically varied in concentration, a psychophysical function can be established. This function should specify how much of a sensory change can be expected with changes in ingredient level. Obviously, ingredients with flatter functions show greater 'recipe tolerance'. The concept of recipe tolerance has applications in manufacturing and quality control. There is a greater liability in manufacturing errors when the dose-response function is steep, and recipe tolerance is low, than when the dose response-function is flat and recipe tolerance is high (given equal impact on overall acceptability). Another application is in cost reduction reformulations in which an ingredient will be reduced in concentration. How much can it be changed before there is a major impact on the sensory properties of the product? Only by careful scaling can this be known. Gathering such scaling data in the early stages of product development can potentially save time later.

The creative aspects of the practice of product innovation, unfortunately, mitigate against these applications of scaling. There is a strong temptation to formulate a single candidate product, 'optimized' at the benchtop using the personal opinion of the research chemist. This brainchild is then submitted for evaluation against some competitive standard in a simple-minded two product test. To the extent that a product developer tends to formulate at the benchtop without consumer input, the benefits of systematic variation of ingredients will be lost. To avoid this tendency, the formulator must understand the benefits of experimental design, and be willing to submit *variables* for evaluation (ranges of products) rather than single items for an uninformative paired test versus some standard.

IDENTIFICATION OF IMPORTANT SENSORY ATTRIBUTES

Descriptive analysis is the generic term for procedures which attempt to make analytical specifications of the sensory attributes of a product and their perceived intensities. In practice, this usually involves some degree of panelist screening and training, and arrival at a consensus for the meanings of the terms to be used. A critical aspect of any descriptive analysis procedure is the process of identification of those critical terms and attributes. This problem can determine the direction of a descriptive program in its early phases. However, the process of identifying attributes has received remarkably little attention in the evaluation literature, and almost no research into alternative approaches.

In practice, the establishment of the set of terms to be used in a given evaluation has several common themes. The attribute generation process may involve a literature review. The panelists themselves will almost always have input into the choice of terms, acting as if they were pseudo-consumers. In some cases, the panel leader will choose the critical terms, based on knowledge of the physical attributes of the products, the sensory modalities involved and existing methods. Some group discussion follows to eliminate redundant terms and clarify the meanings of others. Physical reference standards may be provided whose sensory characteristics exemplify what is meant by that term. Finally, some statistics may be brought to bear. Multivariate techniques such as factor analysis or principal components can provide two main directions. Correlations among terms can be assessed and aid in the elimination of redundant items. In reality, the elimination of a term on the basis of its high correlation with other items is rare. It is much easier to add terms to the ballot, especially if the product has been researched for a period of time, than it is to remove terms. Finally, multivariate statistics can also build models which allow some insights into which attributes appear most highly correlated with consumer acceptance, certainly a good reason for retaining them as part of the descriptive system.

Several criteria have been proposed for defining what is a good or useful term in descriptive analysis (Harper *et al.,* 1968; Civille & Lawless, 1986). A simple requirement is that the sensory term be a recurring and salient attribute of that product class. If the products are generally salty, saltiness should be on the ballot. Second, the term should be capable of differentiating at least some of the products in the class. Saltiness becomes a little less interesting if all products have a uniformly high level of saltiness as opposed to the case where there is *variation* in saltiness.

If the set of terms as a whole is appropriate, it should be possible for a person to read the sensory profile of a product and then identify that product in a multiple choice test from among a set of similar items. This type of backwards analysis task has been used as a check on the accuracy of a description. It has also been used in wine research to assess the degree of communication among judges (Lehrer, 1983; Lawless, 1984). From an applied perspective, a good descriptive analysis program for a given product class will be useful in product development, optimization, correlation with consumer preferences, correlation with instrumental measures, in quality control and reformulation efforts (Civille & Lawless, 1986).

If some of the above criteria have been achieved and applications appear to be running smoothly, a sensory evaluation practitioner will often conclude that he or she has developed a satisfactory set of descriptive terms. This stands in stark contrast to the degree of intellectual agonizing in the psychological literature over so-called 'basic' tastes, primary colors, and such (Ishii & O'Mahony, 1987). The goal of such research, historically, has been to establish a set of atomistic nonreducible terms, that are not composites of simpler sensations, and that are mutually exclusive and exhaustive of the sensory qualities in that modality. Sensory evaluation practitioners might profit from some of this intellectual confrontation. There is great temptation, for example, to use integrated consumer language on a descriptive ballot. For example, a complex adjective like 'light' or 'fresh' can have multiple sensory aspects determining panelists' reactions. This is in contrast to simple terms which could be more closely related to physical references and physical ingredients of a product, and thus be more actionable to the formulating chemist.

Other psychological techniques might be worth trying in order to upgrade the methods used in attribute identification. One technique from social psychology and personality theory is the repertory grid procedure (Kelly, 1955). Here, products are evaluated in triads. The job of the observer is select the outlier in each triad, and then describe on what basis the outlier is different from the two more similar items. Other triads may be constructed until the set of attributes is exhausted. The set often contains many bipolar items (as in the semantic differential). The technique has recently been applied in food research (McEwan & Thomson, 1988).

Many ballot-generation sessions in descriptive analysis bear a resemblance to focus groups used in qualitative marketing research with consumers. There are principles involved in conducting 'good' qualitative groups, some of which may be applied to the early descriptive analysis sessions. For example, it's considered good form to suppress dominant panelists and encourage passive individuals. The techniques for qualitative sessions have evolved from the interview procedures used in clinical psychology and psychiatry (Goldman &

Schwartz McDonald, 1987). Knowledge of these principles and techniques would expand the technical repertoire of many descriptive panel leaders.

Other techniques will undoubtedly be brought to bear on this critical step in sensory evaluation. In exploring a new product category, sorting tasks may be helpful. In such an approach, subjects are asked to group items on whatever basis they feel is important. The groupings may then be probed to identify critical attributes, much as in the repertory grid procedure, but without being limited to triads as groupings. Recently, such sorting procedures have been used in concert with multidimensional scaling, cluster analysis and attribute ratings to build models of the perceptual categorization of fragrance materials (Lawless, 1988*b*).

Sensory terms may be viewed as learned concepts. A concept or category-learning approach has recently been suggested for perceived qualities in the senses of taste and olfaction (Ishii & O'Mahony, 1987; Lawless, 1988*a*). This approach recognizes that observers *learn* to divide up their sensory experiences into useful groupings. This process of perceptual grouping (and conversely differentiation) lends a developmental perspective to the search for good sensory attributes. That is, the set of useful terms need not be seen as a static entity arrived at by physiological experimentation, but rather as a fluid functional system whose lowest-level or most fundamental terms depends upon the level of perceptual differentiation and language development that the panelists have currently achieved.

This places the specification of sensory terms squarely in the camp of other knowledge-based hierarchies. We learn to divide up our sensory world in ways that are useful. If our life, job, etc., does not demand fine distinctions, then we may have no language and no perceptual mechanisms to make them. This idea may also be useful in describing the differences in the perceptions of trained panelists and consumers. Sensory terms form a hierarchy. Some are more generic (e.g. 'citrus') and some entail more specific distinctions (orange, lemon, lime, etc.). The process of subdivision may proceed to unknown depths as expertise is gained. In addition, terms at different levels may have different perceptual properties. Rosch *et al.* (1976) have expounded the theory of *basic levels* which signify useful levels in which the category is sufficiently differentiated from neighboring categories, but still encompasses a set of partially heterogeneous items. The basic levels in the classification of objects have properties which are different from subordinate or superordinate categories. They are the nouns first learned by children, for example. Consideration of hierarchies and levels might prove beneficial in the definition of sensory attributes, especially in consumer research.

PERCEPTUAL ISSUES: MEMORY AND CONTEXT

A classical model for the definition of perception in psychology has it that

Perception = Sensation + Interpretation

The interpretation process involves some reference to information stored in memory, with which the sensory data are compared. Virtually all so-called sensory judgments involve humans *actively processing* the basic sensory information before making their ratings on a ballot. The idea that a trained panelist can relate to the fundamental atomistic barebones bits of sensory data without being integrative is a throwback to the ideas of classical introspectionism, and an idea that must constantly be questioned. The effects of stored information (i.e. of memory on sensory judgments is not often considered in sensory evaluation and represents a fertile area of research opportunities.

Sensory researchers are becoming aware of memory influences in the recognition of *context effects*. This has a rich history in the study of rating scale behavior, and is discussed from two major perspectives. One viewpoint holds that ratings are biased in a certain manner by influences of the stimulus set as a whole, by the rating scale itself and by the ways in which the two are matched up by observers (e.g. Poulton, 1979). The second orientation proposes that human observers are by their nature relative judgment machines. That is, even single stimuli are judged relative to some frame of reference in memory.

Helson (1964), in his adaptation-level theory, recognized that many previously occurring stimuli would affect the judgment (and rating assigned) to a given stimulus. The physiologically based adaptation effects of immediately preceding stimuli were part of Helson's equation. These effects have been well studied in the sense of taste, for example (Bartoshuk *et al.,* 1964; O'Mahony & Godman, 1974). In addition, Helson recognized that more remote stimuli, from earlier in the experimental session and from earlier occasions of exposure to similar stimuli would also have effects. Such effects must proceed through retrieval of stored information, since sensory impressions or sensory images of these stimuli are no longer present.

The viewpoint that sensory ratings are a relative judgment process, as opposed to an absolute or invariant process, has received theoretical attention in the work of Parducci (1965) (see also Anderson, 1977). Parducci's range–frequency theory captured two very strong human predispositions in their use of rating scales. One tendency is that over many stimuli or products, humans will tend to distribute their ratings across the available scale, which is largely true except for certain avoidances of extreme response categories. The second tendency is that people like to use the scale points an equal number of times, that is they don't like to *overuse* a certain response alternative.

This last tendency manifests itself in any task in which the response alternatives consist of fixed categories. Imagine a professor who gave four-alternative multiple choice tests in which the correct answer was always item 'C'. Many students would become more and more uncomfortable as the test progressed with continuing to answer 'C'. This tendency also may explain why panels of employees that are involved in acceptability testing tend to be more critical than outside consumers. People who see nearly optimized products on many occasions may get tired of rating everything at the top of the scale, and begin to push their judgments down. The equal-frequency principle has been well documented for multiple-stimuli sessions, especially on its influence on psychophysical functions (Riskey *et al.*, 1979; Lawless, 1983).

Sensory professionals must appreciate the human as a relative judgment machine. After all, people do not generate numbers in the same way that a physical assessment device like a pH meter does. Instead, we act as devices which constantly recalibrate themselves. The simplest manifestation of this process is in contrast effects. A 40 degree (F) thaw seems wonderfully mild and pleasant after a 20 below zero Wisconsin cold spell, but the same 40 degree temperature seems chilly after a summer heat wave. Thus sensory-based judgments can be pushed around on the rating scale, depending upon what recent stimuli they contrast with.

The implications of this behavior for sensory evaluation are profound. The author once boasted to a vice-president of research and development that he could produce any rating he wanted for a product (given the opportunity to 'adjust' the context). The educational part of his message to him was that he should not view ratings as absolute measurements. The product that received a mean value of 5·5 in last week's testing might in fact be better or worse than the product that got a 5·5 this week, depending upon the other items that were seen at about the same time.

The second implication for applied sensory procedures concerns trained and 'calibrated' descriptive panelists. It is widely thought that descriptive panelists can be anchored to physical reference standards that represent certain sensation levels on a scale. Thus, after sufficient training, they will behave like physical analytical instruments and produce reliable absolute judgments. Whether or not the biasing effects of context can be eliminated or even attenuated by descriptive analysis training is open to question.

A third implication concerns the opportunity for new evaluation procedures that take advantage of these relative judgment tendencies, rather than viewing them as biases to be eliminated, trained away, etc. An alternative approach to scaling is to make all ratings relative to some standard product, instead of pretending that ratings have absolute meaning. Rated degree of difference from control has been rarely applied in sensory work, although overall degree of difference was recently proposed as an approach to difference testing in heterogeneous product systems, as a replacement for the traditional forced-choice procedures (Aust *et al.*, 1985).

Such procedures would seem to have potential for stabilizing judgments in single-attribute ratings as well as in overall degree of difference. Furthermore, they remove some of the variation and problems with individual differences in sensitivity that plague the attempts to force-fit panelists into uniformity. Due to specific anosmia or taste blindnesses or other influences, what is 'strong' on a scale to one panelist, may in fact be 'weak' to another, considering their overall levels of the intensity of sensory experiences. Attempting to eliminate this difference by saying, 'OK, that reference may be weak to you, but rate it at the top of the scale anyhow', seems questionable. Rating *relative* to a reference removes this problem. Insofar as a source of unwanted variation (i.e. a panelist-by-product interaction) is eliminated, test sensitivity may be improved. Attribute ratings can then be correlated with acceptability or appropriateness ratings for those attributes, for example, on 'just-right' scales.

These effects concern the interactions of multiple stimuli for a single attribute. We can also investigate the interactions of multiple rating scales for a single stimulus. Are there some sensory attributes which seem to have an influence on one another? At a physiological level, there are well documented effects of mixture components on one another (Lawless, 1986), in which intensity ratings for a given stimulus or attribute are raised or lowered by the presence of other physical stimuli presented at the same time. We can also ask about perceptual effects, i.e. those not clearly related to physiological mechanisms. In realizing that humans are integrative perception devices, we expect interactions.

As mentioned above, a simple example of an attribute interaction is in the sour/sweet ratio which determines the acceptability of many fruit-derived products like wine. The high acidity of German riesling seems most acceptable when balanced with residual sugar. In a product optimization process, it would be fruitless, so to speak, to try and examine these sensory characteristics one variable at a time. They must be optimized simultaneously.

A largely undocumented area of inquiry involves the interactions of attributes across modalities. For example, it is widely believed that fragrance attributes can influence the perceived functional characteristics of consumer products. It is not unreasonable to think that a detergent with a fresh, clean lemon smell might seem more efficacious than the same chemical formula without the supporting fragrance. Documenting these trends in consumer perception would seem to be a promising area of investigation, especially in fragrance work. This also raises questions as to the origin of such associations in the consumer's mind. Are they dependent upon childhood experiences? Are they a function of adult experiences with products, foods, cleaners and such? Certainly there are trends in the consumer marketplace in the choice of fragrance themes for particular categories of products like detergents and cleaners — these choices are not arbitrary.

It is also widely recognized, in the unspoken model discussed above in scaling, that certain sensory attributes will drive the overall acceptability of a product. But what about the influence of *the act* of making attribute ratings? If I have rated a product very positively when it comes to sensory attributes and functional characteristics (e.g. cleans well, dries fast) will I be more prone to give the product a high acceptability rating at the end of the questionnaire, than if I have not made those positively weighted attribute ratings? Has my attention been drawn to aspects of the product that I would not have considered very important in terms of my recollections of the product as a whole? Conversely, if the overall opinion question is at the beginning of a questionnaire or interview, will I then go and seek reasons to justify my acceptability rating in the more detailed questions that follow? Such biases are expected in consumer research but rarely documented.

Memory issues

A cardinal rule of questionnaire design has always been to ask *only* about that which the respondent has readily available to recall. 'How many times did you wax your floor in the last 5 years' leads more often to bewilderment than to reliable responses. In spite of this warning in consumer research, little attention has been paid to the question of ratings of stimuli that are not presented in the experimental session, but are given an intensity rating after some delay between presentation of the product and evaluation. How does the evaluation of a product change under conditions of recall versus actual observation?

Barker and Weaver (1983) studied this question for recalled images of taste and smell stimuli, and found that people tend to underestimate the intensity of recalled stimuli. That is, when asked to match a physical stimulus to what they recalled a previous stimulus to be, they chose a physical stimulus that was weaker, i.e. lower in concentration than the original stimulus. Given the poverty of our imaging capabilities, this is not surprising. One can remember the exhilaration of skydiving, but there's nothing like the real experience. In apparent contradiction to the report of Barker and Weaver, Osaka (1987) reported that scaling the *remembered* concentrations of pyridine yielded a steeper psychophysical function than when pyridine was actually smelled. While the matching and scaling tasks are very different and a potential methodological source of this apparent contradiction, further research is needed to explore the issue.

The importance of memory processes in consumer research is clear when one considers that products are usually not compared side-by-side in real life. Homemakers will buy one yogurt this week and another brand next week. How good is their memory for remembered flavor? A provocative study by Glaeser and Riesterer (1985) left 24 h between yogurt evaluations and found reliable preference rankings in about 25% of subjects. As this is less than would be expected in side-by-side presentation, one might conclude that delay has an effect, one that is deleterious to either discrimination of preference.

Another reliable effect appears to be that the task can be made harder or easier, as one might expect, depending upon the physical separation of the stimuli. Results from reaction time experiments for relative size judgments of remembered stimuli indicate that the discrimination becomes harder as the physical difference of the stimuli decreases (Moyer, 1973). As one would expect, judging the difference between a remembered elephant and a remembered mouse is easier than judging the difference between a remembered husky and a remembered wolf. As in the case of the overt psychophysics of physical stimuli, there is some disagreement as to whether the memory scale has ordinal or interval properties (Banks *et al.*, 1982).

The functional delay between presentations of products has implications for analytical sensory tests as well as for consumer research. In the case of products with taste and smell properties, in which adaptation could play an important part, one might expect a short delay to enhance performance, insofar as adaptation would be minimized and recovery enhanced. This would presumably happen at some cost to the accuracy of the memory of the first-presented stimulus. The tradeoff function has been difficult to establish. Frijters (1977) addressed this question for the triangle test and found little or no effect. Again, an area for further research.

METHODS RESEARCH

One hallmark of a genuine technology is the visible effort expended at improvements in methods. If sensory evaluation is to emerge as a legitimate scientific discipline, resources must be directed both in academic and industrial settings at experiments on new techniques. Head-to-head comparisons of methods, on the same products and with the same observers have been rare in the sensory literature. Furthermore, investigators have often become bogged down in pseudo-theoretical issues such as what type of psychophysical function fits the data. This is of limited relevance to evaluation professionals who are dealing with minor product improvements, small formula changes and cost reduction efforts a large part of the time. For example, most models of psychophysical functions cannot be differentiated as to goodness-of-fit when the range of product variation is in the one or two JND range. What criteria, then, can we bring to bear on any comparison of methods? Several themes from psychological testing are germane, notably the concepts of reliability and validity.

Reliability describes the degree to which the measurement instrument gives you the same answer upon multiple observations. Obviously, this idea has several different applications in practice. One can ask whether repeating the evaluation (i.e. the whole test) using the same products but with a different random sample of consumers yields the same product differences and same mean score values. One can also ask whether repeating the same test with the same observers gives the same differences and mean values. Agreement can also be addressed among a panel of judges. That is, a related issue concerns whether the panel, especially when trained, is homogeneous. What is the degree of 'error' variance due to panelist differences and panelist by product interactions?

One popular approach to this question has been the examination of panelist by product interactions in analysis of variance tables. This is a sensitive measure when other sources of variance are low, but has some liabilities. If significant, it is a clear indicator of inter-judge disagreement. One could attribute such a finding to lack of training, or to misunderstanding of descriptors and scales, of reference standards or of anchors which represent certain scale points. On the other hand, there could be legitimate physiologically based differences in sensitivity or responsiveness to some families of stimuli, as in the case of specific anosmia.

The converse situation is even more risky. In finding no panelist-by-product interaction, one cannot conclude that panelists are necessarily homogeneous. Failure to reject the null hypothesis is ambiguous. The F ratio for the panelist–product interaction includes some estimate of mean squares due to panelist–by–product systematic variance *divided by* some error term, usually some higher interaction terms such as panelists–by–products–by–replicates. Thus the size of this F ratio and its statistical significance is as much affected by panelist–product variance as by the error term. A statistically nonsignificant result means only that the *relative* size of the panelist–product mean squares was small when compared to the error. If the higher interaction term has greater mean squares, the null will fail to be rejected. However, panelist agreement could still be terrible!

Correlational techniques may be an alternative to this ambiguous approach (see for example, Brien *et al.*, 1987), if not as a replacement for the analysis of variance then at least as an additional analysis. Different correlations can also be assessed to capture different aspects of inter- and intra-judge agreement. Do judges agree with themselves (for a given set of products) from session to session? Do they agree with other panelists? From session to session as well as within a session? Does the panel as a whole reproduce its mean scores when a set of products is repeated? Do they drift from day to day? All of these questions may need to be answered separately, depending upon the nature of the panel and products and the need for replicates within a session and across sessions.

The validity question is also important, although somewhat harder to deal with experimentally. Does the evaluation technique measure what you intend it to measure? The validity of a single attribute or related set of attributes is one issue. One can ask whether a set of theoretically related attributes shows correlated values across products. For example, one might expect the terms clean and fresh to be correlated in the consumer's mind, and partially related to the overall acceptability of a product. Such correlational techniques are closely related to the idea of construct validity in psychological testing. That is, if you think you're tapping into some underlying construct in the consumer's mind, then a set of hypothetically related attribute scales should product a pattern of intercorrelation in a factor analysis, for example.

Another validity question concerns the relationships among evaluation techniques or among panels. The results from an internal or employee panel may or may not predict the responses of consumers from outside the R&D environment. In screening fragrance candidates for consumer products, the first data to be gathered outside of the supplier's or perfumer's evaluations may be employee panels. The highest scoring items from this screening tool may go to further development and testing. If the instrument is to have demonstrated its predictive validity, however, some of the fragrances which score poorly should also be carried forward to discover, in fact, *whether they fail* in the consumer environment. Since industries are loathe to expend resources on testing poorer products, this correlation is rarely assessed. The fact that a candidate fragrance wins in both situations is necessary but not sufficient to demonstrate predictive validity. Other candidates which scored low in early screenings and have dropped by the wayside might have done just as well in further testing as the early-round 'winners'.

Correlation between hedonic assessments is only one form of predictive validity. One can also ask whether the data from a descriptive panel, which strictly speaking should not be involved in hedonic estimates, can be related in mathematical models to the acceptability estimates of consumers. Presuming that your modeling capabilities are adequate, the construction of a successful regression equation, for example, goes far to establish that the descriptive panel is measuring sensory attributes that mean something in real life. In addition, some degree of correspondence to instrumental measures is also useful. Validity can be approached from the viewpoint of establishing the convergence or interconnectedness of various evaluations.

These concepts are related to the ideas of precision and accuracy in the physical sciences. Precision has to do with the error surrounding a given measurement as does reliability. Accuracy has more to do with meeting some otherwise 'true' or criterion value in a measurement, a value that is established by other means. The classic metaphor for precision and accuracy is in archery. An archer who is precise but not accurate will have a tight grouping of arrows, but away from the bull. The accurate but not precise archer will have a loose grouping, but one that centers on the bull. This is the classical situation in sensory evaluation, in which we hope that the mean panel scores will say something meaningful, and we live with a certain amount of error variance. Lack of precision is then dealt with statistically.

Another important criterion for the utility of a sensory evaluation method is its sensitivity to product differences. One can argue that sensitivity to differences can only be achieved with an instrument which is reliable. However, it may be worth examining this more global property in its own right. Finding small product differences is often a challenge for evaluation professionals who are supporting research efforts in a class of products where optimization is nearly complete.

In recent head-to-head comparisons of scaling methods, such an approach was taken using t values and F ratios for product differences as estimates of the signal-to-noise ratio for a scaling method (Lawless & Malone, 1986). This is reasonable since these statistical measures are estimates of scaled product differences divided by an estimate of error variance. One could ask whether each method yielded significant differences or not, but this reduces the comparison to a binary yes/no situation. In contrast, test statistics provide a more continuous measure of the instrument's sensitivity.

There is good theoretical justification for using test statistics as signal-to-noise measures. In a comparison of the reliabilities and errors in measurement in the physical and behavioral sciences, Hedges (1987) used differences between treatment means divided by standard errors as a means of comparisons on unitless dimensions. The psychological scaling measures used in signal detection theory and Thurstonian scaling are also derived from differences between means divided by estimates of dispersion (Frijters, 1979a). In all these cases, experimental error is used as yardstick, either to estimate the resolving power of an evaluation procedure, to estimate the reliability of a whole series of measurements, as Hedges has done, or to estimate scale values for perceived degree of differences, as in the indirect scaling methods from signal detection and Thurstonian scaling.

From a broader perspective, one can look at an improvement in evaluation procedures as any change in method or analysis that reduces error variance (conversely, one that improves reliability or precision). A friend who served in the Coast Guard once told the author about painting the ship from stem to stern and then starting over again, to fight 'Old Man Rust'. Rust was always with them, and it was a never-ending battle. Unexplained error variance is the 'Old Man Rust' of sensory (and all behavioral) research. Obvious approaches to reducing error variance include tighter experimental control and more precise psychophysical techniques, but practitioners also need to think about statistical partitioning of formerly unexplained variation. The gist of this approach is to remove some variance from the error term, thus improving the signal-to-noise ratio. Since inter-individual variation is one major source of error in sensory procedures, approaches which attempt to explain, reduce or partition variance due to systematic individual differences will improve the resolving power of an evaluation procedure.

Several sources of individual differences deserve greater attention from evaluation professionals. First, the literature on physiologically-based differences in sensitivity is continuing to expand. Two examples are the specific anosmias to particular chemical families of olfactory stimuli (Amoore, 1977) and the loss in responsiveness to oral trigeminal irritants among individuals with high levels of red pepper consumption (Lawless *et al.*, 1985). Another source of individual variation in sensory performance concerns personality differences that are related to the ways in which stimuli are interpreted in the process of rating, and how additional stimuli present in the context of the sensory evaluation (e.g. product claims) may influence some consumers' expectations and judgments (Stevens *et al.*, 1988).

Finally, the whole area of perceptual learning needs to be considered in assessing the ways in which sensory judgments change as a function of training and or expertise (e.g. Lawless, 1984). This is needed for two reasons. First, an understanding of the process of perceptual sharpening, especially in olfaction, should aid in the development of improved descriptive panel training methods. Second, an understanding of the ways in which perceptions change as a function of training would more carefully establish the potential differences in perception between experts and consumers. Although this has obtained occasional attention in some fields like wine evaluation (e.g. Lehrer, 1983), it is widely assumed that the perceptual apparatus of experts reflects only a more discriminative level of resolving power with an associated enhancement in qualitative vocabulary. It is also possible, however, that training creates fundamental alterations in the peripheral sensory mechanisms (e.g. Freeman, 1983) such that the very basic incoming sensory data, upon which the higher perceptual apparatus operates, are qualitatively altered. If so, it's no surprise that relating the judgments of experts to consumers turns out to be a difficult task.

REFERENCES

Amoore, J. E. (1977). Specific anosmia and the concept of primary odors. *Chem. Senses,* **2,** 267–81.

Anderson, N. H. (1977). Note on functional measurement and data analysis. *Percept. Psychophys,* **21,** 205–15.

Aust, L. B., Gacula, M. C., Beard, S. A. & Washam, R. W. (1985). Degree of differences test method in sensory evaluation of heterogeneous product types. *J. Food Sci.,* **50,** 511–13.

Banks, W. P., Mermelstein, R. & Yu, H. K. (1982). Discriminations among perceptual and symbolic stimuli. *Memory Cogn.,* **10,** 265–78.

Barker, L. M. & Weaver, C. A. (1983). Rapid, permanent loss of memory for absolute intensity of taste and smell. *Bull. Psychon. Soc.,* **21,** 281–4.

Bartoshuk, L. M., McBurney, D. H. & Pfaffmann, C. (1964). Taste of sodium chloride solutions after adaptation to sodium chloride: implications for the water taste. *Science,* **143,** 967–8.

Bressan, L. P. & Behling, R. W. (1977). The selection and training of judges for discrimination testing. *Food Technol.,* **31,** 62–7.

Brien, C. J., May, P. & Mayo, O. (1987). Analysis of judge performance in wine quality evaluations. *J. Food Sci.,* **52,** 1273–9.

Civille, G. V. & Lawless, H. T. (1986). The importance of language in describing perceptions. *J. Sensory Stud.,* **1,** 203–15.

Ennis, D. M. & Mullen, K. (1985). The effect of dimensionality on results from the triangular method. *Chem. Senses,* **10,** 605–8.

Freeman, W. J. (1983). The physiological basis of mental images. *Biol. Psychiat.,* **18,** 1107–25.

Frijters, J. E. R. (1977). The effect of duration of intervals between olfactory stimuli in the triangular method. *Chem. Senses,* **2,** 301–11.

Frijters, J. E. R. (1979a). Variations of the triangular method and the relationship of its unidimensional probabilistic models to three-alternative forced-choice signal detection theory models. *Brit. J. Math. Stat. Psychol.* **32,** 229–41.

Frijters, J. E. R. (1979b). The paradox of disciminatory nondiscriminators resolved. *Chem. Senses.,* **4**, 355–8.
Glaeser, H. & Riesterer, M. (1985). Untersuchengen zum sensorischen Gedachtnis des Konsumenten. *Alimenta,* **24**, 3–6.
Goldman, A. E. & Schwartz McDonald, S. (1987). *The Group Depth Interview.* Prentice Hall, Englewood Cliffs, NJ.
Green, D. M. & Swets, J. A. (1966). *Signal Detection Theory.* John Wiley, New York.
Harper, R., Bate Smith, E. C., & Land, D. G. (1968). *Odour Description and Odour Classification: A Multidisciplinary Examination.* American Elsevier, New York.
Hedges, L. V. (1987). How hard is hard science, how soft is soft science? *Am. Psychol.,* **42**, 443–55.
Helm, E. & Trolle, B. (1946). Selection of a taste panel. *Wallerstein Lab. Comm.,* **9**, 181–94.
Helson, H. (1964). *Adaptation-Level Theory.* Harper & Row, New York.
Ishii, R. & O'Mahony, M. (1987). Taste sorting and naming: can taste concepts be misrepresented by traditional psychophysical labelling systems? *Chem. Senses,* **12**, 37–51.
Kelly, A. T. (1955). *The Psychology of Personal Constructs.* W. W. Norton, New York.
Lawless, H. T. (1983). Contextual effects in category ratings. *J. Test. Eval.,* **11** 346–9.
Lawless, H. T. (1984). Flavor description of white wine by 'expert' and non-expert wine consumers. *J. Food Sci.,* **49**, 120–3.
Lawless, H. T. (1986). Sensory interactions in mixtures. *J. Sens. Stud.,* **1**, 259–74.
Lawless, H. T. (1988a). Odour description and odour classification revisited. In *Food Acceptance,* ed. D. M. H. Thomson. Elsevier Applied Science Publishers, London.
Lawless, H. T. (1988b). Categorization of ambiguous odors in restricted and unrestricted classification tasks with multidimensional scaling analysis, presented at Association for Chemoreception Sciences, Sarasota, FL.
Lawless, H. T. & Malone, G. J. (1986). A comparison of rating scales: sensitivity, replicates and relative measurement, *J. Sens. Stud.,* **1**, 155–74.
Lawless, H. T. & Schlegel, M. P. (1984). Direct and indirect scaling of sensory differences in simple taste and odor mixtures. *J. Food Sci.,* **49**, 44–6, 51.
Lawless, H., Rozin, P. & Shenker, J. (1985). Effects of capsicin on gustatory, olfactory and irritant sensations and flavor identification in humans who regularly or rarely consume chili pepper. *Chem. Senses,* **10**, 579–89.
Lehrer, A. (1983). *Wine and Conversation.* Indiana University Press, Bloomington.
Marks, L. E. (1974). On scales of sensation: prolegomena to any future psychophysics that will be able to come forth as a science. *Percept. Psychophys.,* **16**, 358–76.
McBride, R. L. (1983). A JND/category scale convergence in taste. *Percept. Psychophys.,* **34**, 77–83.

McBride, R. L. (1987). Taste psychophysics and Beidler equation. *Chem. Senses,* **12,** 323-32.
McEwan, J. A. & Thomson, D. M. H. (1988). An investigation of the factors influencing consumer acceptance of chocolate confectionery using the repertory grid method. In *Food Acceptance,* ed. D. M. H. Thomson. Elsevier Applied Science Publishers, London.
Moskowitz, H. R. (1983). *Product Testing and Sensory Evaluation of Foods.* Food and Nutrition Press, Westport, CT.
Moyer, R. S. (1973). Comparing objects in memory: evidence suggesting an internal psychophysics. *Percept. Psychophys.,* **13,** 180-4.
O'Mahony, M. & Godman, L. (1974). The effect of interstimulus procedures on salt taste thresholds. *Percept. Psychophys.,* **16,** 459-65.
Osaka, N. (1987). Memory psychophysics for pyridine smell scale. *Bull. Psychonom. Soc.,* **25,** 56-7.
Parducci, A. (1965). Category judgment: a range-frequency model. *Psych. Rev.,* **72,** 407-18.
Poulton, E. C. (1979). Models for biases in judging sensory magnitude. *Psych. Bull.* **86,** 777-803.
Riskey, D. R., Parducci, A. & Beauchamp, G. K. (1979). Effects of context in judgments of sweetness and pleasantness. *Percept. Psychophys.,* **26,** 171-6.
Rosch, E., Mervis, C. B., Gray, W. D., Johnson, D. M. & Boyes-Braem, P. (1976). Basic objects in natural categories. Cog. Psychol., **8,** 382-439.
Stevens, D. A., Dooley, D. A. & Laird, J. D. (1988). Explaining individual differences in flavour perception and food acceptability. In *Food Acceptance,* ed. D. M. H. Thomson. Elsevier Applied Science Publishers, London.

5

Integration Psychophysics

ROBERT L. MCBRIDE
Sensometrics Pty Ltd, 357 Military Road, Mosman, NSW 2088, Australia

&

NORMAN H. ANDERSON
*Department of Psychology, University of California, San Diego,
La Jolla, California 92093, USA*

INTRODUCTION

Perception of flavor involves integration of a number of sensations, both within and between sensory modalities. Within the taste modality alone, sensations of sweet, sour, salty and bitter may operate; the aroma, similarly, may consist of several odor qualities. Moreover, sensory integration is not limited to the chemical senses: the appearance of a food, its textural quality, even the sound of chewing — all can contribute to overall perception. Cognitive factors, such as knowledge and expectations, may also play a role. In eating food, therefore, as in many other areas of perception, the human organism acts as an integrator of stimulus information.

How to analyze this integration is a central problem. Are there simple rules that underlie the various integration processes? Do the components add in their effects? Do they interact? To what extent does each sensory and nonsensory component contribute?

The purpose of this chapter is to present *information integration theory* as a foundation for studying these questions. As a research paradigm, integration theory has been employed in several areas of psychophysics (e.g. Anderson, 1970, 1974a; Massaro & Anderson, 1971; Weiss, 1975; Clavadetscher & Anderson, 1977; Carterette & Anderson, 1979), as well as in general judgment theory (e.g. Anderson, 1974b, 1981). Integration methods can also be useful in the chemical senses, but only a few studies have so far been reported (Klitzner, 1975; McBride, 1982a, 1986, 1989; De Graaf et al., 1987; McBride & Johnson, 1987; McBride & Finlay, 1989). Accordingly, this chapter begins with a brief overview of theory and method, and then discusses a number of applications in the chemical senses.

THE LOGIC OF INFORMATION INTEGRATION THEORY

Conceptual basis

The conceptual basis of integration theory is shown schematically in the integration diagram of Fig. 1. In the first stage, which concerns sensory action, physical stimuli are transduced into psychological sensations. In Fig. 1, the physical stimuli are denoted by S_i and the corresponding psychological sensations by s_i, such that $s_i = V(S_i)$. This *valuation function*, V, corresponds to what is ordinarily called the psychophysical function.

The second stage in Fig. 1 involves integration of the separate sensations, s_i, into an overall, conscious percept, r. Thus, $r = I(s_1, s_2, \ldots)$, where I is the *integration function*.

The third stage is concerned with observable, behavioral reactions to the integrated percept — typically with judgments of flavor intensity or

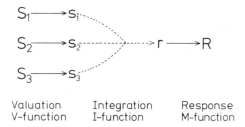

FIG. 1. The logic of information interpretation theory. A chain of three linked functions, V, I and M, connects the observable stimuli, S_i, and the observable response, R. The valuation function, V, maps the physical stimuli, S_i, into sensations, s_i. The integration function, I, maps these individual sensations, s_i, into an integrated sensation, r. The response function, M, maps this covert sensation, r, into the observable response, R. (After Anderson, 1981, p. 5.)

of overall hedonic tone. This step requires a mapping, M, from the unobservable r to the observable response, R. Thus, $R = M(r)$, where M is the *response function*.

Simple though it is, the process model in Fig. 1 represents a marked departure from traditional psychophysics. Traditional psychophysics has focused exclusively on the psychophysical function, whereas the aim of integration analysis is to obtain information on *all three* stages of the integration diagram; that is, on the form of the three functions V, I and M. This might seem impossible given that all three functions are theoretical concepts and not directly observable. Indeed, the only observables are the physical stimuli and the physical response. Nevertheless, from these observables, all three theoretical functions in the integration diagram can be inferred.

In empirical applications of integration theory, a key tool is the analysis of *response patterns*. With two stimulus variables, subjects would be presented as stimulus pairs combined according to a two-way factorial design. For each pair, they would judge total intensity, say, or the intensity of one specified component on a quantitative scale.

These responses would be graphed in the form of a factorial plot. The pattern in the factorial plot can, under certain conditions, elucidate the V, I and M in the integration diagram, as illustrated in the two following experimental examples.

Example 1: Taste–odor integration
To illustrate integration of odor and taste, suppose subjects rate the perceived total intensity of mixtures of sucrose (sweet taste) and orange flavor (orange odor). An obvious hypothesis is that total intensity of the mixture is the sum of the intensities of the two components. In terms of the integration diagram in Fig. 1, this hypothesis may be written

$$r = s_{su} + s_{of} \quad (1)$$

To investigate this hypothesis, nine combinations of sucrose and orange flavor are presented according to a 3 × 3 factorial design. The factorial plot in Fig. 2 shows that both variables affected the judgment of intensity. The upward slopes of the curves represent the effect of increasing the sucrose level; the vertical separation between the curves shows similarly that increasing the concentration of orange flavor also increases the response.

The question is whether these responses were generated by the hypothesized additive rule. The answer is *yes*; the observed pattern of parallelism points to an additive integration.

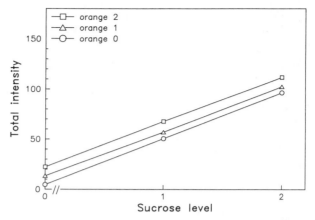

FIG. 2. Factorial plot of perceived total intensity of a smell (orange odor) and a taste (sweetness). The parallelism signifies an additive integration function.

Intuitively, the meaning of parallelism may be seen by considering the spacing between the top and bottom curves. The vertical difference between them shows how much the perceived total intensity is increased by the addition of 0 · 2% w/v orange flavor. This difference is constant, the same for each level of sucrose. Thus, the pattern of parallelism points to an additive integration.

This example is not merely hypothetical. The data in Fig. 2 are real, obtained in a recent experiment on taste–odor integration (McBride, 1988b; unpublished data).

Example 2: Mixture suppression

In mixture suppression, one component of a mixture acts to decrease the perceived intensity of another. In a sugar–acid mixture, for example, stronger concentrations of sugar will decrease the perceived intensity of the acid.

A plausible hypothesis is that mixture suppression will obey a subtraction rule:

$$r = s_{ac} - k s_{su} \qquad (2)$$

where k is a suppression coefficient. There is one complication to this rule. Taste intensity cannot be negative, so the equation must incorporate a threshold, with $s_{ac} > k s_{su}$.

A 4 × 4 factorial plot is shown in Fig. 3. At the three highest acid concentrations, the four curves are parallel, supporting the subtraction

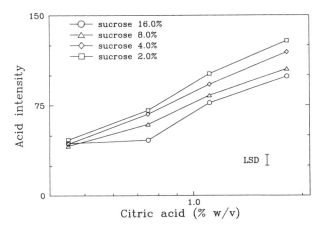

FIG. 3. Factorial plot of perceived acidity of mixtures of sugar and citric acid in lemon drink. The parallelism at all but the lowest acid concentration signifies a subtractive suppression (integration) function.

rule: the vertical difference between the top and bottom curves is the same at all three acid concentrations. This difference shows how much perceived acidity is suppressed by increasing sucrose concentration from 2 to 16%.

At the lowest acid concentration, however, the curves converge. Here the acid concentration is near-threshold, so even the lower sucrose concentrations are sufficient to suppress it.

This example also is not merely hypothetical. It contains real data from McBride & Johnson (1987). Here again, the response pattern in the factorial plot points directly to the integration rule for mixture suppression.

The parallelism theorem

The foregoing discussion shows how the patterns in the factorial graphs of Figs 2 and 3 provide evidence on the integration function, I. These patterns also provide information on the psychophysical function, V, and the response function, M. The logic of these conclusions rests on the *parallelism theorem* (Anderson, 1981, Section 1.2), which will be illustrated here in simple form without proof for the taste–odor integration of Fig. 2.

Suppose the addition rule holds, so that

$$r = s_{su} + s_{of},$$

and that the response scale is linear, so that

$$R = c_1 r + c_0$$

where c and c_0 are constants that represent the unit and zero of the response measure. Then (a) the factorial data plot will form a set of parallel curves and (b) the row means of the factorial data table will be a linear scale of the sensation values of the row stimuli (and similarly for the column stimuli).

Observed parallelism thus provides prima facie evidence for both premises of the theorem. These two premises are taken jointly. Thus, parallelism supports the hypothesis of additive integration. At the same time, it supports the premise that the observable response, R, is a true mirror of the unobservable perception, r. In terms of the integration diagram of Fig. 1, parallelism argues that I is additive and M is linear.

Information on the psychophysical function, V, may also be obtained directly from the factorial graph. Thus, the curves in Fig. 2 define the psychophysical function for sucrose (albeit at only three points in this illustration). Since these curves are parallel, their mean curve has the same shape. But the mean curve is based on the column means of the factorial graph, which, by part (b) of the theorem, provides a true linear scale of the subjective intensity values of the column (sucrose) stimuli.

The concentrations of sucrose in the experiment were in fact logarithmically spaced, so the psychophysical function for sucrose is logarithmic. Similarly, the parallel curves in Fig. 3 indicate that the psychophysical function for perceived acidity is also approximately logarithmic. Each of these functions, it should be emphasized, was that actually operative in the mixture task itself.

Observed parallelism, it should be recognized, is a powerful empirical result. If one premise of the theorem is true and the other false, the observed data will not be parallel. Thus, if the integration were additive but the response function nonlinear, the data would show nonparallelism. Similarly, if the response function were linear but the integration nonadditive, then also the data would not be parallel.

Parallelism cannot, of course, *prove* that both premises are true. Logically, parallelism could still be obtained if the integration were nonadditive and the response function were nonlinear. But this would require that the nonlinearity in the response exactly offset and cancel the nonadditivity in the integration. This logical possibility may need serious consideration as an alternative interpretation of parallelism in some situations. By ordinary scientific standards, however, it seems fair to say that observed parallelism constitutes strong evidence for the two premises, additivity and linearity, taken together.

To sum up, observed parallelism can provide evidence about all three theoretical functions in the integration diagram: the psychophysical function, V; the addition (subtraction) rule, I; and the response function, M. The evidence is empirical, not a mathematical proof.

Although the finer details of integration analysis cannot be considered here, one point needs to be made explicit: the premise of response linearity is not actually necessary. In its general form, the parallelism theorem requires only the additivity assumption itself; no assumption about the response scale is needed. An empirical illustration is given by Carterette and Anderson (1979) and the general theory and supporting statistical methods are given in Anderson (1982). In this general form of the parallelism theorem, all that is at issue is the algebraic structure of the integration rule and that is empirically testable.

Experimental procedure

To have a linear response scale is extremely helpful in practice. In the initial work on integration analysis, however, there was no clear way to select such a scale. The two most common numerical response measures, rating and magnitude estimation, both had strong traditions in psychophysics, so both were employed in integration experiments. On the whole, rating produced parallelism whereas magnitude estimation did not. The accumulated results thus indicated that rating has better claim to be the linear response measure.

Magnitude estimation, as various writers have remarked, appears to be biased by diminishing returns in number usage, so that 110 and 111 are less different than 10 and 11. To express equal sensory differences therefore requires greater numerical differences at higher intensities. This means that the response numbers themselves are not a linear function of sensations. No way to control this bias has been presented.

The rating method is also subject to certain biases, but fairly simple procedures have been developed that seem to largely eliminate them (see Anderson, 1982, Chapter 1). The essential requirement is that the subject develops a stable frame of reference to relate the range of stimuli and the assigned response range. One useful way is to begin with end-anchors, stimuli just noticeably more extreme than the range of experimental stimuli, which anchor the end points on the rating scale and help set up the frame of reference. Preliminary practice with these end-anchors and intermediate stimuli can eliminate the well-known stimulus distribution effects, as well as certain end biases in the response scale. Graphic, line-mark format is considered best because it avoids residual preferences that can appear with category-type format. Continuing attention to rating procedure is important to improve present methods. Possession of a linear response method may be almost essential for analysis of stimulus interactions and configurality.

CONCEPTUAL ADVANTAGES OF INFORMATION INTEGRATION THEORY

The central advantage of information integration theory is that one simple experiment can simultaneously provide information on three distinct aspects of sensory processing: (1) the integration rule; (2) the psychophysical functions for the component stimuli; and (3) the validity of the response measure. No preliminary scaling of the separate component stimuli is needed; no statistical transformations are required. All this information is available merely by inspecting the factorial plot of the data. Other conceptual advantages of particular relevance to the chemical senses also deserve mention.

Mixture analysis
In the chemical senses, stimulus mixtures are of basic interest from both theoretical and applied standpoints. Traditional psychophysics, however, has largely been concerned with psychophysical functions of single, unmixed stimuli. This is understandable, indeed logical, if determination of psychophysical functions of individual stimuli is considered prerequisite to the study of their behavior in a mixture. But such single-variable analysis is not satisfactory, in part because it fails to provide a validity criterion for the psychophysical function, in part because the psychophysical functions of the individual stimuli bear no necessary relation to mixture psychophysics.

In integration theory, the psychophysical functions emerge from the structural rules governing the integration of the mixture components. Mixture rules, not psychophysical functions, are primary. This was clearly demonstrated by the foregoing experiment in which taste suppression occurred at the integration stage. The experiment investigated suppression directly without recourse to individual psychophysical functions. Indeed, the nature of the psychophysical functions is almost irrelevant to the integration rule.

Nonmetric stimuli
Psychological dimensions do not always have underlying physical correlates. Acceptability varies among foods, yet there is no purely physical measure that underlies this variation; texture, likewise, lacks a simple physical correlate. For food science, though, these complex stimuli are far more important than the one-dimensional stimuli commonly studied in the psychophysical laboratory.

Complex stimuli can be handled by information integration theory. No prior metric is required, either physical or psychological; the parallelism analysis can measure complex stimuli on a validated, linear scale. Cognitive factors (food labeling, advertising, attitudes) which play a part in determining overall hedonic response can be handled in exactly the same way as the sensory factors (color, flavor). The valuation function, V, is thus more general than the psychophysical function. It does not require a physical metric, and therefore allows analysis to be extended beyond the limits of traditional psychophysics.

Other integration rules
The methodology of integration theory can be applied to other rules besides the adding-type rules. For multiplication rules, there is a linear fan theorem exactly analogous to the foregoing parallelism theorem (Anderson, 1981, Section 1.4). Averaging rules are useful for measuring importance weight separately from scale value and for comparing qualitatively different variables (Anderson, 1974*a*, p. 226, 1982, Section 2.3.2).

Of special interest are rules that involve stimulus interaction. Two of these — the subtractive threshold rule (Fig. 3) and the dominant component rule (to be discussed later; Fig. 5) — have a mathematical structure that can be diagnosed from the pattern in the factorial graph. In general, however, interaction may not correspond to any simple mathematical structure. But if the response is known to be linear, the observed pattern in the factorial graph mirrors the underlying perception and so can provide useful information on stimulus interaction. This is a major reason for concern with developing methods and procedures that yield linear response measures.

TECHNICAL ADVANTAGE OF INFORMATION INTEGRATION THEORY

Traditional methodology in the chemical senses has been handicapped by a number of technical difficulties. Three fairly common difficulties will be explained here, and it will be shown how they can be resolved with integration psychophysics.

Predicting the sum from judgments of the components

Suppose we wish to test simple additivity in binary mixtures — for example, do taste intensity and odor intensity add together to give overall flavor intensity? In the traditional test, subjects judge the intensity of the individual stimuli and of their mixture. Judgments of the individual (unmixed) stimuli are then added and compared against the judgment of their mixture.

This test seems simple and straightforward, but it relies on strong assumptions: the judgments of the individual stimuli, and of their mixture, must be made on a common ratio scale. In other words, the response measure must be a genuine linear scale, with the same unit *and* a true zero point in all three cases. Few if any published studies have shown these conditions to be satisfied.

The first problem concerns the linear response measure. Without a validity criterion such as parallelism, there is no way of assessing whether a response measure is linear. This is essential to the test of additivity.

The second problem concerns the common unit. In psychophysical scaling, there is a tendency for subjects to spread their responses over a constant response range, regardless of the actual stimulus range presented (see Poulton, 1979; Anderson, 1982, Sections 1.1, 3.12, 3.13; McBride, 1982*b*, 1985). For example, an odor stimulus with a small subjective range would tend to evoke the same range of overt responses as a taste stimulus with a large subjective range. If odor, taste and mixture stimuli are judged at three different sessions, the scale unit must be expected to vary. The two sets of individual judgments would not have the same unit; adding them together would make no more sense than adding inches and centimeters. To ensure a common unit experimentally, it is generally necessary to obtain all three sets of judgments in a single session (Anderson, 1982, Section 1.1.8).

Third, a true zero point is also necessary. To illustrate, suppose that in a mixture experiment all judgments have been obtained on a linear response scale. Each judgment is therefore equal to its true sensation value except for an added constant, c_0. In the traditional check on additivity, this constant will be added *twice* when the individual components are summed, but only once in the judgment of the mixture (Anderson, 1982, pp. 313–14). Even if additivity actually held, the traditional check would deem it to fail.

All these problems raise doubts about interpreting the traditional check on additivity. Cases of apparent nonadditivity that have been reported in the literature may merely reflect artifacts in the analysis.

These problems, however, are bypassed with the methodology of information integration theory, because separate judgments of the individual stimuli are not necessary. The test of additivity can be made directly on the response pattern of the factorial design.

For some integration tasks, this feature is indispensable. Suppose, for example, we wish to study the hedonic integration of flavor and temperature. It would be meaningless to present the temperature dimension in isolation; it is only when the subject is presented with various flavor levels, *at* various temperatures, that the task becomes meaningful.

Multiple regression analysis

Most of the problems in the preceding section can also be avoided by using standard multiple regression — or so it would seem. In multiple regression, the simple sum is replaced by a weighted sum plus an additive constant. The weights adjust for scale units, and the additive constant adjusts for zero points. These adjustments would allow individual stimuli and their mixtures to be judged at separate sessions, if so desired.

But regression analysis has pitfalls of its own (see also following subsection). One concerns the typical use of the *physical* values of the stimulus predictors. These physical values cannot give a proper test of the integration model, for which *subjective* values are essential. In contrast, the integration test of additivity automatically allows for subjective values.

Regression analysis has a further pitfall: the regression weight for each stimulus variable is confounded with the scale unit for that variable. Hence, the weights do not, in fact, measure relative importance, as has often been assumed. This problem cannot be overcome by standardization of the weights (see Anderson, 1982, Sections 4.3 and 6.1).

Weak inference

The scatterplot is often used to check additivity: insensity judgments of mixtures are plotted against the intensities predicted by adding the judgments of the separate components. High coefficients are taken as support for additivity.

But correlation and scatterplot statistics are 'weak inference' methods; they seem to test the integration model but do not really do so. Many empirical illustrations are given by Anderson (1982, Chapter 4). In a study of length bisection, for example, the additive model yielded a correlation coefficient of $0 \cdot 9997$, yet the data showed significant, meaningful nonadditivity.

In contrast, the factorial plots used in integration theory are graphically more informative and can pinpoint deviations from additivity. Statistically, moreover, the hypothesis of additivity is equivalent to zero interaction in the analysis of variance. This standard statistical method provides a strong test of the additive model. The factorial plot also displays the relative contributions of the individual components, which cannot be seen from a scatterplot. In Fig. 2, for example, sucrose contributes more to total intensity than orange flavor.

Prescribed integration rules

A class of tasks that has been useful in other areas of psychophysical integration may be called prescribed tasks: subjects are instructed to judge the sum, average or difference of two stimuli. The main attraction of prescribed tasks is for measurement. The instructions prescribe additivity, so it is not surprising when parallelism appears in the data. But parallelism would not appear unless the response measure were a true linear scale. In judging average grayness of two Munsell chips, for example, parallelism was obtained with ratings but marked non-parallelism with magnitude estimation (see Anderson, 1982, Fig. 1.3). This outcome supports the linearity of the rating response. Furthermore, linear scales of the stimulus values are obtainable by virtue of the second conclusion of the parallelism theorem. Prescribed tasks thus offer a potentially easy way to handle the two problems of measurement, response and stimulus.

The difference task arises naturally in preferences between two choice alternatives. An application to food preferences is given by Shanteau and Anderson (1969) and two applications to taste preferences are cited in the following paragraph. The sum and average tasks can allow for more than two stimuli, an advantage over the difference task. All three tasks can be used to study adaptation and other order effects in terms of serial integration. A final advantage of prescribed tasks is to study stimulus interaction and configurality. Thus subjects could be asked to judge the sum or differences of two mixtures that do not themselves obey any simple integration rule. Stimulus values of the mixtures could nevertheless be obtained from the prescribed task, and these values could then be used to shed light on the mixture interaction.

Two taste studies have used a prescribed integration task: one on sweet–bitter integration by Klitzner (1975), discussed later, and one on glucose–fructose mixtures by De Graaf *et al.* (1987). As De Graaf *et al.* emphasize, parallelism analysis can provide psychological sweetness scales of the mixtures themselves. This can provide information on the preconscious sensory integration, even when this does not follow any simple algebraic rule. Of added interest was the use of a water stimulus as a potential means of providing a zero point and hence a ratio scale. Although prescribed tasks seem artificial, these examples indicate they can help analyze complex integration processes that may not follow simple rules.

EMPIRICAL STUDIES OF INTEGRATION

The usefulness of information integration theory in the chemical senses will depend heavily on the operation of adding rules or other simple rules of stimulus integration. A number of promising studies have been published. Some have not been specifically concerned with integration rules, and others suffer from lack of statistical power; several are open to some question about the validity of the response measure. These matters will not be considered in the following survey, whose main aim is to point up some attractive areas of further work.

Appearance-flavor

DuBose *et al.* (1980) studied the combined effects of colorant and flavor in both cake and beverages. Most relevant here is one set of data which showed that addition of yellow colorant increased the perceived intensity of the lemon flavor of cake. Color and flavor were varied in a factorial design, and the resulting factorial plot was approximately parallel (DuBose *et al.*, 1980, Fig. 8). Although the authors did not discuss the matter, this observed parallelism suggests that the integration of flavor and color obey an adding-type rule.

Appearance–flavor studies hold special interest because, with different sensory pathways, possible nonadditivities from peripheral interaction are excluded. The integration of the two stimuli must be central, therefore the adding-type models seem not unlikely. Moreover, the effects themselves can be substantial, not only for color on taste (Pangborn, 1960; Johnson et al., 1983), but also for color on aroma (Christensen, 1983).

Taste–odor
Figure 2 illustrated a taste–odor integration that followed an adding rule. This agrees with a previous study (Hornung & Enns, 1984), in which subjects sipped instant coffee from the lower of two cylindrical compartments while smelling a second solution of instant coffee from the upper compartment. Intensities of taste and odor could thus be manipulated independently with this device.

In one condition, the taste and odor components were varied in a factorial design, with subjects required to judge the overall intensity of the combined stimulus. The factorial plot was roughly parallel and the interaction statistically nonsignificant, supporting the hypothesis of an adding rule (cf. also Enns & Hornung, 1985; Hornung & Enns, 1986).

Two other studies of taste–odor interaction have reported 'near-misses' to additivity, with the overall intensity of the mixtures somewhat less than the sum of the intensities of the unmixed components (Murphy et al., 1977; Murphy & Cain, 1980). However, these studies utilized the traditional method of predicting from the sum of the components. This method, as noted above, depends on assumptions of common unit and true zero. Although attempts were made to satisfy these assumptions, there is no way to check their validity.

Homogeneous taste mixtures
Mixtures of similar substances, such as different sugars, are possibly integrated at some sensory level and so might not obey any simple integration rule. In an investigation of the sweetness of binary sugar mixtures (McBride, 1986), the factorial plots deviated from parallelism at high concentration, as illustrated in Fig. 4. The convergence pattern implies that sweetness is subadditive at higher intensities. This interpretation has been supported by a scale-free, sweetness matching experiment (McBride, 1988a).

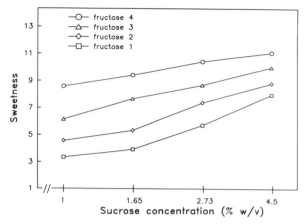

FIG. 4. Factorial plot of the sweetness of mixtures of sucrose and fructose. The curves converge in the upper-right of the plot indicating a subadditive integration (after McBride, 1986).

This subadditivity is consistent with electrophysiological data. Jakinovich (1982) found that the integrated *chorda tympani* response of the gerbil to mixtures of sucrose and saccharin followed the equation

$$R_{mix} = R_A + R_B - (R_A \cdot R_B / \max R_{mix}) \qquad (3)$$

where R_A and R_B are the neural responses to sweeteners A and B, respectively; R_{mix} is the response to their mixture; and max R_{mix} is the maximum neural response to a mixture of A and B. In physiological terms, this model represents the 'interaction of two substances with two independent receptor sites through a common effector system which limits the maximum response' (Jakinovich, 1982, p. 50). When either R_A or R_B is small, the subtractive term would be negligible and the simple adding model would effectively apply. As R_A and R_B increase, however, additivity would fail, yielding a convergence pattern like that of Fig. 4.

Heterogeneous taste mixtures: perception of overall intensity

Heterogeneous taste mixtures are those in which at least two components are perceptible (e.g. sweet-acid). The overall perceived intensity of these mixtures has generally been reported to be less than the sum of the perceived intensities of the components.

Recent integration studies of perceived total intensity of sugar–acid mixtures (McBride, 1989; McBride & Finlay, 1989) suggest the operation of a *dominant component model*: the subjectively dominant component by itself determines the total intensity of the mixture; there is no integration of the components at all.

The response pattern implied by a dominant component model is evident in the factorial data plot of Fig. 5 (from McBride, 1988b). Consider the mean intensity rating for the top level of citric acid alone (leftmost data point of top curve). As increasing concentrations of sucrose are added to this level of citric acid, that is, moving rightwards along the curve, there is no increase in total intensity at all. Even the addition of a high (0·80M) sucrose concentration does nothing to augment the total intensity of the mixture. This pattern also holds for the second highest level of citric acid, with the possible exception of the mixture on the extreme right of the curve.

The middle concentration of citric acid, likewise, dictates the total intensity of the mixture, *until* the sucrose concentration reaches 0·37M; at this point and at the next (0·80M), the total intensity of the mixture is almost the same as the total intensity of the sucrose when tasted alone (bottom curve). At these two points, the sucrose alone dictates the intensity of the mixture.

The curve second from the bottom displays the same 'switchover'

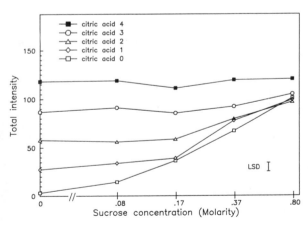

FIG. 5. Factorial plot of the total intensity of mixtures of sucrose and citric acid. The response pattern conforms to a dominant component model (see text; after McBride, 1989).

pattern: total intensity is dictated by citric acid at the three lowest levels of sucrose, but by sucrose at the two highest levels of sucrose. Very similar patterns have been reported in another recent integration study of sucrose–citric acid and sucrose–sodium chloride mixtures (Frank & Archambo, 1986), and also for sucrose–citric acid in lemon drink (McBride & Johnson, 1987).

Heterogeneous mixtures: perception of components

Perception of components in a heterogeneous taste mixture, sweet and acid, was shown earlier in Fig. 3. There was good support for a subtractive model of mixture suppression.

This same pattern was obtained in a more extensive study of sugar–acid mixtures (McBride, 1988*b*). Figure 6 gives the factorial plots for sweetness (A) and perceived acidity (B). The curves are essentially parallel at the top three abscissa concentrations in each plot, supporting a subtractive model. Convergence at the lower concentrations can be accounted for by the threshold adjustment mentioned earlier.

Lawless (1977, Table 1) varied sucrose and quinine sulfate in a factorial design, with subjects required to judge sweetness and bitterness on open-ended graphic rating scales. A plot of these data is given in Fig. 7. Again, the same pattern is evident in both parts: the curves are approximately parallel in the upper right, but converge as they approach the abscissa. This constitutes further support for suppression as subtraction and demonstrates again the simplicity of response pattern analysis — the mere factorial plot of the data tells all.

Odor mixtures

Data on olfactory integration are meager, so postulation of integration models is necessarily speculative. However, there is some indication (Laing & Willcox, 1983, Fig. 3) that the dominant component model may apply to the total intensity of heterogeneous odor mixtures. Also, the perception of components in heterogeneous odor mixtures may, as for taste mixtures, follow a subtractive model. Laing *et al.* (1984) varied binary combinations of six qualitative different odorants in factorial designs, using graphic rating as the response measure. For higher odor intensities, judgments of the components followed a pattern similar to that in Fig. 6. For lower odor intensities, however, the data did not seem to exhibit the same subtractive pattern, a finding documented elsewhere (Lawless, 1977, Tables 3 and 4).

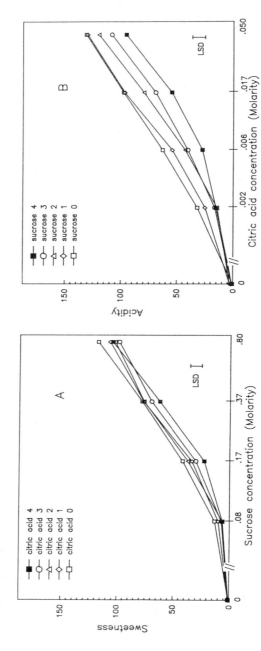

FIG. 6. Factorial plots of the sweetness (A) and acidity (B) of mixtures of sucrose and citric acid (after McBride, 1989).

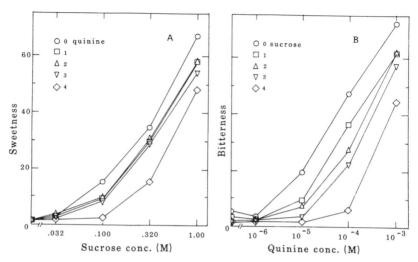

FIG. 7. Factorial plots of the sweetness (A) and bitterness (B) of mixtures of sucrose and quinine sulfate (after Lawless, 1977, Table 1).

Hedonic integration

The acceptability of a food — its integrated hedonic tone — largely determines its sale and consumption. Studies of hedonic integration show qualified support for adding-type models. Factorial designs were used in the following studies, and the data showed approximate parallelism in several cases. Still, there were also some deviations from parallelism whose interpretation is uncertain.

DuBose et al. (1980) found support for an adding model in the hedonic integration of fruit drinks. Flavor and color were varied independently, and the factorial plots exhibited near-parallelism, both for cherry drinks and orange drinks. In contrast, the hedonic ratings for cake displayed a crossover interaction: increasing the color intensity could either increase or decrease acceptability, depending on the flavor level.

Klitzner (1975) applied information integration theory to study the pleasantness of sweet-bitter mixtures of apple juice and quinine sulfate. The factorial plot showed nonparallelism, with the quinine curves converging at higher concentrations of apple juice. This could reflect an averaging (as distinct from adding) process, with greater weighting of higher quinine concentrations, or merely a nonlinearity in the response scale.

To resolve this ambiguity, Klitzner performed a second experiment that involved two integration operations. Subjects were required to judge the *difference* in pleasantness between two mixtures. Although the hypothesized greater weighting might affect the first integration, which governed the pleasantness of each mixture, it presumably would not affect the second integration, which governed the difference judgment. As it turned out, the difference judgments yielded parallelism, implying that the nonparallelism in the first test was not substantive and could be attributed to response bias. This study is only suggestive, but it does demonstrate the potential of the two operation analysis (Anderson, 1982, Section 5.6) as a means of resolving uncertainty about response linearity.

Lawless (1977) obtained pleasantness ratings of mixtures of sucrose and quinine sulfate, using both sip-and-spit and dorsal flow procedures, and also of two odor mixtures, pyridine and lavender oil, and heptanal and lemon oil. Data from the sip-and-spit procedure are approximately parallel (Lawless, 1977, Table 2). This additive pattern thus supports the finding of Klitzner. In contrast, the data for odor pleasantness (Lawless, 1977, Tables 3 and 4) show substantial deviations from parallelism.

In an applied sensory evaluation study, McBride and Faragher (1978) investigated the effect of a hormonal spray on the acceptability of apples. There were three levels of spray (including zero) and three

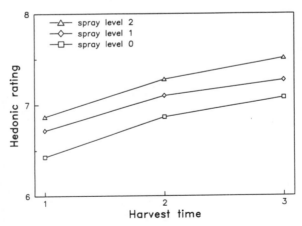

FIG. 8. Factorial plot of the acceptability of apples harvested at three stages and treated with three levels of spray. Parallelism signifies an additive integration function (after McBride & Faragher, 1978).

harvest times, constituting a 3 × 3 factorial design. A taste panel scored the acceptability of the apples on a nine-point hedonic rating scale.

The results are summarized in Fig. 8. The apples improved with later harvest, and also with the spray treatment. Most important, from the integration perspective, the curves are parallel. To reiterate the parallelism principle, this pattern suggests (a) the effect of the spray was simply to 'add' acceptability to the apples, (b) the increase in acceptability with harvest time (the 'psychophysical function') is almost linear, and (c) the overt responses on the hedonic scale are valid representations of the subjective acceptability sensation.

INTEGRATION PSYCHOPHYSICS — A CONCEPTUAL SHIFT

Mechanisms of stimulus integration have long been of concern in the chemical senses, but little is known about the exact nature of the various integration functions. The empirical data reviewed in this article, limited though they are, suggest that a number of common integration tasks obey simple algebraic models. A more solid empirical foundation is needed, but the available data seem promising.

Fundamentally, integration psychophysics represent a conceptual departure from traditional psychophysics. Traditional psychophysics centered on the question of the psychophysical law: the relation between psychological sensation and physical stimulus intensity. The crux of this question lies in the measurement of sensation, but this is not resolvable within the limits of single-variable psychophysics.

The essence of the conceptual shift is the focus on the many-variable integration function, rather than the single-variable psychophysical function — hence the lable *integration psychophysics*. Mathematically, the many-variable integration function provides constraints that can resolve the measurement problem. Substantively, sensory measurement and sensory processing become dual aspects of a single operation.

ACKNOWLEDGEMENTS

The basis for this chapter was prepared while the first author was a Visiting Scholar in the Center for Human Information Processing, Department of Psychology, University of California, San Diego, 1984. This work was supported in part by Grant BNS82-07541 from the

National Science Foundation to the Center for Human Information Processing. We thank W. S. Cain and L. E. Marks for their comments on an earlier draft.

REFERENCES

Anderson, N. H. (1970). Functional measurement and psychophysical judgment. *Psychol. Rev.*, **77**, 153–70.

Anderson, N. H. (1974*a*). Algebraic models in perception. In *Handbook of Perception,* Vol II, ed. E. C. Carterette & M. P. Friedman. Academic Press, New York, pp. 215–98.

Anderson, N. H. (1974*b*). Information integration theory: a brief survey. In *Contemporary Developments in Mathematical Psychology,* Vol II, ed. D. H. Krantz, R. C. Atkinson, R. D. Luce & P. Suppes. Freeman, San Francisco, pp. 236–305.

Anderson, N. H. (1981). *Foundations of Information Integration Theory.* Academic Press, New York.

Anderson, N. H. (1982). *Methods of Information Integration Theory.* Academic Press, New York.

Carterette, E. C. & Anderson, N. H. (1979). Bisection of loudness. *Percept. Psychophys.*, **26**, 265–80.

Christensen, C. M. (1983). Effects of color on aroma, flavor and texture judgments of foods. *J. Food Sci.*, **48**, 787–90.

Clavadetscher, J. E. & Anderson, N. H. (1977). Comparative judgment: tests of two theories using the Baldwin figure. *J. Exp. Psychol.: Human Perception and Performance*, **3**, 119–35.

De Graaf, C., Frijters, J. E. R. & van Trijp, H. C. M. (1987). Taste interaction between glucose and fructose assessed by functional measurement. *Percept. Psychophys.*, **41**, 383–92.

DuBose, C. N. Cardello, A. V. & Maller, O. (1980). Effects of colorants and flavorants on identification, perceived flavor intensity, and hedonic quality of fruit-flavored beverages and cake. *J. Food Sci.*, **45**, 1393–9, 1415.

Enns, M. P. & Hornung, D. E. (1985). Contributions of smell and taste to overall intensity. *Chem. Senses*, **10**, 357–66.

Frank, R. A. & Archambo, G. (1986). Intensity and hedonic judgments of taste mixtures: an information integration analysis. *Chem. Senses*, **11**, 427–38.

Hornung, D. E. & Enns, M. P. (1984). The independence and integration of olfaction and taste. *Chem. Senses*, **9**, 97–106.

Hornung, D. E. & Enns, M. P. (1986). The contributions of smell and taste to overall intensity: A model. *Percept. Psychophys.*, **39**, 385–91.

Jakinovich, W. (1982). Stimulation of the gerbil's gustatory receptors by saccharin. *J. Neurosci.*, **2**, 49–56.

Johnson, J. L., Dzendolet, E. & Clydesdale, F. (1983). Psychophysical relationship between sweetness and redness in strawberry-flavored drinks. *J. Food Prot.*, **46**, 21–5.

Klitzner, M. D. (1975). Hedonic integration: test of a linear model. *Percept. Psychophys.*, **18**, 49–54.

Laing, D. G. & Willcox, M. E. (1983). Perception of components in binary odour mixtures. *Chem. Senses*, **7**, 249–64.

Laing, D. G., Panhuber, H., Willcox, M. E. & Pittman, E. (1984). Quality and intensity of binary odor mixtures. *Physiol. Behav.*, **33**, 309–19.

Lawless, H. T. (1977). The pleasantness of mixtures in taste and olfaction. *Sens. Process.*, **1**, 227–37.

Massaro, D. W. & Anderson, N. H. (1971). Judgmental model of the Ebbinghaus illusion. *J. Exp. Psychol.*, **89**, 147–51.

McBride, R. L. (1982a). Toward a Unified Theory of Psychophysics. Unpublished PhD thesis, Macquarie University, Sydney.

McBride, R. L. (1982b). Range bias in sensory evaluation. *J. Food Technol.*, **17**, 405–10.

McBride, R. L. (1985). Stimulus range influences intensity and hedonic ratings of flavour. *Appetite*, **6**, 125–31.

McBride, R. L. (1986). Sweetness of binary mixtures of sucrose, fructose, and glucose. *J. Exp. Psychol.: Human Perception and Performance*, **12**, 584–91.

McBride, R. L. (1988a). Taste reception of binary sugar mixtures: psychophysical comparison of two models. *Percept. Psychophys.*, **44**, 167–71.

McBride, R. L. (1988b). Sensory evaluation: a link between market research and R & D. *Proc. 17th Nat. Conf. Market. Res. Soc. Aust.*, paper 22.

McBride, R. L. (1989). Three models for taste mixtures. In *Perception of Complex Smells and Tastes,* ed. D. G. Laing, W. S. Cain, R. L. McBride & B. W. Ache. Academic Press, Sydney.

McBride, R. L. & Faragher, J. D. (1978). The sensory evaluation of Jonathan and Delicious apples treated with the growth regulator Ethephon. *J. Sci. Food Agric.*, **29**, 465–70.

McBride, R. L. & Finlay, D. C. (1989). Perception of taste mixtures by experienced and novice assessors. *J. Sensory Studies* in press.

McBride, R. L. & Johnson, R. L. (1987). Perception of sugar–acid mixtures in lemon juice drink. *Int. J. Food Sci. Technol.*, **22**, 399–408.

Murphy, C. & Cain, W. S. (1980). Taste and olfaction: independence vs interaction. *Physiol. Behav.*, **24**, 601–5.

Murphy, C., Cain, W. S. & Bartoshuk, L. M. (1977). Mutual action of taste and olfaction. *Sens. Process.*, **1**, 204–11.

Pangborn, R. M. (1960). Influence of color on the discrimination of sweetness. *Am. J. Psychol.*, **73**, 229–38.

Poulton, E. C. (1979). Models for biases in judging sensory magnitude. *Psychol. Bull.*, **86**, 777–803.

Shanteau, J. C. & Anderson, N. H. (1969). Test of a conflict model for preference judgment. *J. Math. Psychol.*, **6**, 312–25.

Weiss, D. J. (1975). Quantifying private events: a functional measurement analysis of equisection. *Percept. Psychophys.*, **17**, 351–7.

6

Cognitive Aspects of Difference Testing and Descriptive Analysis: Criterion Variation and Concept Formation

MICHAEL O'MAHONY
Department of Food Science and Technology, University of California, Davis, California 95616, USA

INTRODUCTION

Much of sensory evaluation involves measurement in three main areas: scaling, difference testing and descriptive analysis. This chapter will not deal with scaling although it is an area that has been the subject of intense study. Rather, the intention is to explore cognitive aspects of the other two main branches of sensory evaluation. For difference testing, the problems of criterion variation will be considered while for descriptive analysis, the mechanisms of concept formation will be examined.

SENSORY DIFFERENCE TESTS: FORCED CHOICE PROCEDURES

Sensory difference tests are behavioral procedures used to determine whether a given judge can distinguish between two similar stimuli. The two stimuli are generally very similar, otherwise there would be no question as to whether they were different and there would be no point in performing any formal tests to investigate the point.

For the sensory evaluation of food, the two stimuli are generally two very similar samples of food. They may be distinguished in terms of all or of some of their sensory characteristics. Such measures are used to determine whether slight changes occur due to product reformulation or processing changes. They may be used to detect slight flavor variations caused by storage or packaging changes. They may be used to guide a product developer towards an ideal or a rival products set of sensory characteristics.

There are several procedures used for sensory difference testing. These are described in several texts (Amerine *et al.*, 1965; Larmond, 1977; Gatchalian, 1981; Frijters, 1984; Jellinek, 1985; Stone & Sidel, 1985; Meilgaard *et al.*, 1987*a*), while several studies have compared their discriminating ability (Helm & Trolle, 1946; Dawson & Dochterman, 1951; Byer & Abrams, 1953; Hopkins, 1954; Pfaffmann *et al.*, 1954; Gridgeman, 1955, 1956; Hopkins & Gridgeman, 1955; Filipello, 1956; Mitchell, 1956; Hogue & Briant, 1957; Dawson *et al.*, 1963; Grim & Goldblith, 1965; Wasserman & Talley, 1969; Spencer, 1979; Frijters, 1981; Pokorný *et al.*, 1981; Frijters *et al.*, 1982; O'Mahony & Odbert, 1985; O'Mahony & Goldstein, 1986; O'Mahony *et al.*, 1986; Buchanan *et al.*, 1987). They nearly all share in common the feature that the judge must demonstrate that he can distinguish between the foods under question; his word that he thinks they are different is not enough. A commonly used procedure is the paired comparison test.

The paired comparison procedure requires a judge to indicate which is the more intense of two stimuli on a given continuum. Thus, he is asked to indicate the 'sweeter' of two food samples or the 'saltier', 'crunchier' or 'more aromatic', etc. If he is consistent in doing so correctly over several paired comparisons, it is taken that he can differentiate the two. For the experimenter to assess whether the judge is correct, he must first know which of the samples is physically 'sweeter' (has more sugar, etc.). If this is not known, consistency of response can indicate that the judge is differentiating the samples. In this test, as in others, binomial statistics are used to determine how many correct paired comparison tests are required for the experimenter to believe that the judge's performance is better than mere chance.

Note that the judge is wisely not asked merely to say whether he can distinguish between the two samples or not. This would put too much trust in the judge. Very often the judge's impression of his own ability to distinguish between samples bears little relation to his actual performance. It is not uncommon for a judge to throw up his hands and announce that he cannot distinguish between two foods when his performance indicates that he can do so consistently. By forcing the judge to demonstrate his ability by indicating the target stimulus, the problems of the judge's assessment of his own ability are avoided. For the same reason, the judge is also not allowed to respond that he cannot tell the difference between two samples; he is forced to choose. This avoids the problem of judges who lack confidence giving a 'cannot decide' response. Most sensory difference tests are *forced choice*.

Another commonly used procedure is the duo-trio test (Peryam & Swartz, 1950; Peryam, 1958). Here, two unknown food samples are presented along with a third standard sample which is the same as one of the unknowns; the task is to identify which of the unknowns is the same as the standard. To be able to do so consistently implies that the judge can distinguish between the unknowns. Again, it should be noted that the judge is not merely asked to indicate whether he thinks one of the unknowns is the same as the standard, he has to prove he can tell by indicating the appropriate unknown sample. Again, the task is forced choice because the judge must pick one of the unknown samples; he cannot give a 'don't know' response.

A little used but simple modification of this test is the dual standard test (Peryam & Swartz, 1950; Peryam, 1958). Here, two standards are presented instead of one; one standard is the same as one unknown while the other standard is the same as the other unknown. This test is little used but is simple and can be more discriminating than the duo-trio test (O'Mahony *et al.*, 1986). Judges have also indicated their liking for the test in several studies (O'Mahony *et al.*, 1985*a, b, c*; O'Mahony & Goldstein, 1987*a, b*). Literature on relative discriminability of sensory difference tests has already been cited and it is not the intention to pursue the topic further. Suffice to say that much of the theoretical approach to this area has been concerned with psychological mathematical model building (Frijters, 1979, 1984; Ennis & Mullen, 1985, 1986*a, b*; Kapenga *et al.*, 1987; Ennis *et al.*, 1988; Mullen *et al.*, 1988). As a complementary approach, the author's research group has been investigating relative discriminability in terms of the physiological and cognitive aspects of the order of tasting within a given test; this approach has been called Sequential Sensitivity Analysis (O'Mahony & Odbert, 1985; O'Mahony & Goldstein, 1986, 1987*c*).

The third commonly used procedure is the triangle test (Helm & Trolle, 1946; Peryam & Schwartz, 1950; Peryam, 1958). Here, three stimuli are given to the judge; two are samples from one food, the third is a sample of the second food. The task is to identify the odd sample; if this can be done consistently, the judge is deemed to be able to discriminate between the two foods.

Once again, the judge has to demonstrate that he can tell the difference between the foods and once again the task is forced choice.

The judge is not allowed a 'don't know' response; he must choose one of the three stimuli.

According to the literature already cited, the paired comparison is often a more discriminating test; judges may distinguish between two foods with a paired comparison when they cannot distinguish between the same foods using a triangle or duo-trio test; the reasons for this are a complex mix of physiological and psychological factors (O'Mahony & Goldstein, 1987c). The duo-trio and triangle tests generally have the advantage, however, in that the nature of the difference does not have to be specified. The duo-trio test differs from the triangle test in that the judge is looking for a similarity not a difference. The number of samples in each test is the same but the search program in the brain is different. Incidentally, it is worth noting that in an earlier version of the triangle test (Helm & Trolle, 1946), the nature of the difference was specified and judges were instructed to find the two samples which were the same, not the one which was different.

There are several other procedures for difference testing. The Fallis-Lasagna-Tétreault test is similar to the triangle test except that an odd sample is selected from four samples not three (Fallis et al., 1962); it has generally been used for threshold measurement in a medical context (Furchtgott & Willingham, 1956; Schelling et al., 1965; Topinka & Sova, 1967). Frijters (1984) describes this and the case where four stimuli are divided into two pairs (choose two sets of two stimuli from four) as tetrad tests. The Harris-Kalmus test (Harris & Kalmus, 1949) is a similar test in which four stimuli are chosen from eight; essentially, it involves sorting eight stimuli into two groups of four. It is commonly used in genetic research for determining taste thresholds for phenylthiocarbamide. This procedure has also been named the octad test by Gridgeman (1956). Other similar methods are mentioned. Meilgaard et al. (1987a) describe a two-out-of-five test while Basker (1980) has produced statistical tables to assess chance levels for tests involving the choice for various numbers of stimuli (2 from 5, 3 from 7, etc.). Although tests involving choice from large numbers of stimuli may be statistically desirable, there are other considerations that must be considered before their use. These will be discussed in a later section (p. 125).

Pokorný et al. (1981) developed a test similar to the duo-trio test which they called the tetrade test. Here, instead of having two unknown stimuli, one of which is the same as a standard, there were three unknown stimuli. Either one or two were the same as the standard. This test may be more efficient statistically but for the reasons given below, it is biased cognitively. Another biased test is the 'A-Not A' test (Peryam, 1958). However, before these biases can be discussed, it is necessary to understand the nature of criterion variation.

SENSORY DIFFERENCE TESTS: CRITERION VARIATION

The notion of response bias in a judge caused by criterion variation is embedded in signal detection theory (Green & Swets, 1966) and is essential for an understanding of the cognitive aspects of sensory difference testing. Generally, when a judge is comparing two food stimuli in a sensory difference test, the foods will be very similar; they will be difficult to distinguish. If they were easily distinguishable, no formal difference testing would have been required. In fact, in the case where stimuli are easy to distinguish, criterion variation has an insignificant effect and the following argument does not apply.

The judge will find it hard to determine whether two almost identical stimuli are the same or different. The difference will be very slight and it will be difficult for the judge to determine whether he is detecting a real difference or imagining it. Again, it is well to remember that this only applies to small differences. When the judge is uncertain and trying to answer the question as to whether he really detects a difference, a second question is implied in the judgement: How great a difference must there be before the judge believes the difference is not due to his imagination? This is not a sensory question, it is a cognitive question. The judge's response will depend on his willingness to commit himself and say that the difference is real. In essence, the judge selects a given degree of difference as his criterion for saying that the difference is no longer imaginary and can be reported as real. Again, this criterion of difference is not a sensory function; it is cognitive. It is set arbitrarily and can vary. If the judge is feeling reckless, he will be more willing to say that two stimuli are different; he will have a more lax criterion of difference. Should the judge be feeling cautious, he will be less willing to commit himself to saying that the two stimuli are different; he will have a stricter criterion.

The criterion can vary during the course of an experiment; the judge's opinion as to what is a real and what is an imaginary difference is not at all stable for such fine discriminations. Naturally, this variation must not be allowed to happen. It would mean that a judge could often unwittingly change his mind and report two stimuli as being different when shortly beforehand he had reported them as the same. The judge's ability to distinguish would not have changed, he would merely have changed his criterion. With food samples, a judge might report one set of foods as different and another identical set the same, merely because of criterion variation. This might incorrectly be interpreted as nonhomogeneity among the samples. There are two main approaches for overcoming criterion variation: signal detection measures and forced choice procedures. The former will be discussed first.

SENSORY DIFFERENCE TESTS: CONTROLLING CRITERION VARIATION BY SIGNAL DETECTION PROCEDURES

It is not the purpose of this chapter to discuss details of signal detection theory; there are texts and papers devoted to this (Green & Swets, 1966; McNicol, 1972; O'Mahony, 1988) while modern introductory texts in psychology also contain sections on signal detection (D'Amato, 1970; Coren et al., 1979; Schiffman, 1982; Goldstein, 1984; Matlin, 1988; O'Mahony, 1988). Here, the intention is to indicate how signal detection measurement circumvents the problem of response bias caused by criterion variation.

The strategy in signal detection measurement is to allow the criterion to vary and arrange for the judge to make the same judgements at several criterion levels ranging from strict to lax. There are various methods for doing this but the most economical is to arrange for the judge to make judgements using several criteria simultaneously. In effect, it is like asking the judge whether he would report a difference if he were using a very strict criterion, whether he were using a fairly strict criterion, a fairly lax criterion, a very lax criterion, etc. The wording of the question would not be exactly as above. It translates into asking whether the judge would say whether he is sure that there is a difference, whether he would say there is perhaps a difference, whether the stimuli might perhaps be the same, whether he is sure the stimuli are the same, etc. If the judge reports he is sure there is a difference, he is using a strict criterion; if he reports that there is a difference but he is not sure, he is using a more lax criterion. Such a response scheme with responses graded in terms of sureness has the effect of forcing the judge to use several criteria simultaneously. The number of sureness levels chosen will depend on the situation. Two levels of sureness ('sure' versus 'not sure') giving four response categories ('different sure'; 'different not sure'; 'same not sure'; 'same sure') are adequate (O'Mahony, 1979). Three levels of sureness ('sure'; 'not sure'; 'I don't know but I will guess') giving six response categories, can also be straightforward for judges to handle (O'Mahony, 1972). It is desirable to have at least two levels of sureness (four response categories) and there should not be a 'don't know' category or else the less confident judges will not be forced to push themselves into differentiating the stimuli.

Having obtained a set of responses for different criterion levels for replicate sets of the two food stimuli, it is then possible to calculate an index of how sensitive the judge is to the difference. There are several indices that can be used: d', $P(A)$ (Green & Swets, 1966), and the R index (Brown, 1974); Brown's R index which is coincidental with $P(A)$ when measured by the technique outlined above, is used widely by the food industry.

It is in the calculation of these indices that the problem of criterion variation is overcome. Each computation involves explicitly or implicitly the plotting of a graph (Receiver Operating Characteristic Curve). Each judgement at a given criterion level corresponds to a point on this graph. Thus, making judgements at two sureness levels will give four response categories and four points on the graph. Once the points are joined, a bow-shaped graph is obtained. The indices, d', $P(A)$ and R index, are computed from the curvature of the graph. The more the graph curves upward in a dome-like shape, the larger the indices obtained. It is here that the measures become criterion free. A judge with a given sensitivity to the difference between two foods will always produce the same curve. Although his criteria may vary from experimental session to experimental session, the points obtained at the various criterion levels will all fall on the same curve. Thus, one set of points may be obtained during one session while a different set is obtained in the next session. Yet, when either set of points are joined up, they will produce the same curve with the same curvature and thus the same signal detection indices of degree of difference. It is in this way that the measures are independent of the actual criteria chosen *per se*. It does not matter what criteria are chosen during an experimental session, as long as several are used. There need only be enough to provide sufficient points to produce the curve and measure its curvature. In the next experimental session, different criteria points may be chosen but they will result in the same curve and thus the same value for the sensitivity index. There are variations on this procedure but all use essentially the same device for overcoming the effects of criterion variation. The calculation of the R index (O'Mahony, 1979, 1988) would appear to bear no relationship to plotting a curve at all, but its computation is the equivalent of computing the proportion of area beneath a Receiver Operating Characteristic curve which is itself a function of its curvature.

In summary, one way of avoiding the criterion problem is to add a sureness judgement to a question which on its own would be prone to criterion variation. To ask whether two foods are the same or different is a question which can be biased by criterion variation: 'How different do two foods have to be to be called different?' The question should not be asked. However, if a sureness judgement is added, then the problem of criterion variation can be overcome by computing an index of difference: 'Are these foods the same or different and are you sure or not sure?' This provides a very simple strategy for avoiding criterion problems.

SENSORY DIFFERENCE TESTS: CONTROLLING CRITERION VARIATION BY USING FORCED CHOICE PROCEDURES

The second approach to controlling variation in the criterion is to use a forced choice procedure. Consider first a paired comparison task. Should two stimuli be presented to a judge and the judge asked whether they were the same or different, the question will be biased; it will be prone to criterion variation. How different do the stimuli need to be before they are reported as different?

However, if a directional forced choice question were asked, the problem would be solved. For example, if one stimulus contained more sugar and the judge were asked to indicate the sweeter of the two, the criterion problem would be solved. The criterion question is whether one of the two stimuli is sufficiently sweeter to be reported as sweeter. By using the directional forced choice question, the experimenter is in effect telling the judge that one of the stimuli will be sufficiently sweeter to be called sweeter; all he has to do is indicate which one. The criterion question is, in effect, answered by the experimental situation. The procedure is equivalent to setting the judge's criterion so that he will report differences of this magnitude as different, always assuming he can detect them.

The other common forced choice procedures use the same strategy. The triangle test arranges for the judge to pick up the odd sample; the criterion question for this test is whether any sample is sufficiently different from the other two to be reported as being different. By informing the judge that one sample will be different from the other two effectively answers the question. All the judge need do is indicate the odd sample. Again, this is equivalent to setting the criterion to that level of strictness that is just sufficient to differentiate one of the stimuli.

For the duo–trio test, when the judge is choosing which of the unknown stimuli are the same as the standard, the implicit criterion question is whether one of the two unknown stimuli are sufficiently similar to the standard to be reported as such. By informing the judge that one of the unknowns will be different, the criterion question is effectively answered and the judge sets his criterion accordingly. Again, the same is true with the dual standard test. By informing the judge that one unknown will match one standard and the other unknown will match the other standard, the experimenter is once again answering any criterion questions beforehand.

Thus, the common forced choice procedures have the useful property that they are criterion-free measures. Unfortunately, this cannot be said for the 'A–Not A' test, which is still recommended in many quarters. The procedure requires a judge to state whether a given unknown stimulus is the same as a standard stimulus or not. Again, the criterion question is whether it is similar enough to be reported as the same. The response will vary as the criterion changes arbitrarily. There is no forced choice comparison to get round the problem. The test is ill conceived and will suffer from response bias if it is used for fine judgements. Either a forced choice comparison must be introduced, in which case, it becomes a duo–trio test, or sureness judgements should be added so that a signal detection index can be calculated (e.g. 'same as A sure', 'same as A not sure', 'different from A not sure', 'different from A sure').

Thus, when one considers the cognitive aspects of testing such as the criterion the judge sets for himself, it is possible to see that not all the common difference tests are free from response bias caused by the criterion's variation. Using a forced choice procedure would seem a simple way of countering criterion variation. Yet, on closer examination, it can be seen that the situation is more complex.

SENSORY DIFFERENCE TESTS: A FORCED CHOICE PROCEDURE ITSELF IS NOT SUFFICIENT TO CONTROL CRITERION VARIATION

For a difference test, the essential strategy for avoiding criterion problems is to avoid *single stimulus judgements*. For a single stimulus judgement, a judge must consider a single stimulus and ask the question whether it is sufficiently different (or similar) to be reported as different (or similar). This question is prone to response bias caused by criterion variation. The 'A-not A' test involves such a single stimulus judgement; to avoid criterion problems here, a sureness judgement can be added. The paired comparison, triangle, duo-trio and dual standard tests involve comparative forced choice judgements. These have the effect of forcing the judge to set his criterion to a sufficient level of strictness or laxity so that he can place the appropriate number of stimuli in each class (e.g. one odd, two the same). However, this forced choice strategy itself is a necessary but not a sufficient condition for controlling criterion variation. Two further conditions must be fulfilled. Firstly, the judge must know the number of unknown stimuli in each class. Secondly, the judge must have sufficiently few unknowns to be able to make comparisons between them all. It is as well to consider these both carefully. If they are not fulfilled, the judge will be forced to revert to single stimulus judgements and so introduce problems of criterion variation into the procedure.

The judge must know the number of stimuli in each class. For the triangle test, the judge knows that only one sample will be odd and can set his criterion accordingly. He knows there are two in one class of stimuli and one in the other. On the other hand, if the judge were told that maybe one stimulus will be odd and maybe not (the numbers in each class may be 2 and 1, or they may be 3 and 0) then the judge must revert to criterion prone single stimulus judgements. He must ask the criterion question of each stimulus: 'Is this sufficiently different from the other two to be regarded as different?'

For the duo-trio test, the judge is told that one of the unknown stimuli will be the same as the standard, one will not; he knows the number in each class. Were he told that maybe one, both, or none of the unknowns, may be the same as the standard, he has once again to revert to single stimulus judgements. He must ask the criterion question of each unknown, whether it is sufficiently similar to the standard to be reported as the same. The same argument applies to the dual standard test. The judge is told that one unknown is the same as the first standard and the other is the same as the second; he knows the number in each class.

With Pokorný *et al.*'s (1981) tetrade test, the judge is given three unknowns. He is told that one will be the same as a fourth standard stimulus and two will not. So far, he knows the number in each class. However, he is also told that it is possible that two may be the same as the standard and one not. Now, maybe one or maybe two will be the same as the standard. Again, the judge has to revert to single stimulus judgements; he must make a separate decision as to whether each stimulus is similar enough to the standard to be reported as the same. Even if the judge decided that one of the two first unknowns was the same and the other different, the third unknown would still require a single stimulus judgement to decide which class it belonged to. Thus, the test is flawed because it allows response bias due to criterion variation. However, should the test be modified so that the judge knew for certain the number of unknowns that were the same as the standard, then criterion problems would be eliminated. A careful reading of Pokorný *et al.*'s (1981) paper indicates an ambiguity. It is just possible from the wording of the paper that the judge might be told the number in each class beforehand, although a straightforward reading certainly would not suggest this. It is unfortunate that the authors were not clear on so vital a point.

The second condition with the forced choice procedure is that there must not be too many stimuli in each class. If there are, a comparative forced choice decision cannot be made and the judge will need to revert to single stimulus judgements.

Imagine a paired comparison where the judge has to divide the two stimuli into two classes: more sweet and less sweet. Because there are only two stimuli to compare, the sensory input from each can be remembered sufficiently for a true comparative judgement to be made and for the criterion to be adjusted so that it is strict or lax enough to differentiate the two. A duo-trio or dual standard test could be seen as a paired comparison with one or two standards provided and again there are few enough unknown samples for a comparative judgement to be made. The triangle test has three unknown samples, yet judges seem able to make a comparison between them. All of these tests may require judges to retaste the food samples because of memory loss but the amount of retasting would seem to be comparatively little.

The paired comparison considered had two samples, one a sweeter sample, one less sweet. Now, imagine that there were twenty of the less sweet samples and twenty of the sweeter samples given in a random array. The judge would be told to sort them into two groups of twenty. He would thus know the number in each class but could he really make a comparative forced choice decision? He would not be able to remember all forty tastes to be able to make a true comparative judgement. The amount of retasting of samples required would be clearly out of the question. He would have to resort to single stimulus judgements. He would need to taste each sample individually and decide whether it was sweet enough to call 'sweeter' and so be placed in that class. He would have to resort to criterion prone single stimulus judgements.

One tactic that might be adopted in such a situation would be to present two standards and give the unknown stimuli in pairs, one of each pair matched to each standard. In this way, the forty samples could be sorted into two groups of twenty. However, this no longer remains a sorting task; it becomes a set of twenty dual standard tests. Another strategy would be to let the judge taste the stimuli successively, using single stimulus judgements, but add a sureness judgement to each, to allow a signal detection index to be calculated.

Should the stimuli be visual and able to be viewed simultaneously, then comparative judgements could be made. The problem of numbers of stimuli only applies to stimuli that must be given in succession: taste, smell, etc. Further, it is well to remember again that this argument applies only to the judgement of fine differences; for large easily detectable differences, criterion variation is an insignificant effect.

So, it can be argued that if, in a forced choice test, the number of stimuli in each class is too great, comparative judgements cannot be made because the judge cannot remember the stimuli; criterion prone single stimulus judgements become necessary. Although the commonly used force choice sensory difference tests do not seem at risk here, the question becomes one of how many stimuli are too much?

Unfortunately, there is no answer because at the time of writing there has been little research into this area. Further, the answer will probably depend on the food in question and the skills and memory of the judge. It is well to remember that here we are discussing the question of too many stimuli in terms of sufficient memory span to make comparative judgements; there are, of course, other factors to consider which depend on the stimuli at hand: adaptation, fatigue, etc.

All that can be said at the time of writing is that forced choice tests which exceed the commonly used ones in their number of stimuli should be approached with caution. The Harris–Kalmus or octad test requires eight stimuli to be sorted into two groups of four; it is possible that this is too many. Picking one stimulus from four (Fallis et al., 1962) or two from five (Meilgaard et al., 1987a) might be satisfactory or might not. Some of the forced choice tasks for which Basker (1980) supplied the statistical tables may have too many stimuli, e.g. pick two from five, three from seven, four from nine, etc. At this time, one can only speculate. The investigation into the necessary information processing and required memory spans has not been performed. Thus, a limit to the number of stimuli used and prior information given to the judge regarding the number of stimuli in each class, supply important riders to the conditions necessary for a criterion free forced choice design. Without them, the forced choice strategy would not overcome criterion problems.

DESCRIPTIVE ANALYSIS: CONCEPT FORMATION

We now leave the general area of sensory difference testing and consider descriptive analysis. It is necessary for those working in the food or personal products industries to be able to communicate to each other the sensations they are experiencing. For this, a common language is needed; judges must agree on the labels to be given to each sensation that is to be communicated.

However, this is not enough. It is important for judges to agree on the range of sensations to be included under that label. For example, what are the many shades of color that should be labelled red? What are the many taste sensations that should come under the heading 'sweet'? It is necessary that the judges' concepts of these sensations should correspond. The judges need to have a commonly held concept for redness and for sweetness. Then, they will include the same stimuli within the target concept and exclude others. This is necessary for precise communication. So, in the problem of communicating sensations, language merely supplies a label for a given concept, the tricky part is getting all the judges to have the same set of concepts. At this point, it is worth examining the cognitive or 'software' processes in the brain that relate language and perception (Miller & Johnson-Laird, 1976); in particular, it is worth focusing on mechanisms of concept formation.

According to accepted theory (Hull, 1920; Pikas, 1966; Nelson, 1974; Millward, 1980), there are two parts to the formation of a sensory concept: abstraction and generalization. As an example, we will consider the sensory concept of 'redness'. For the first part, the concept of redness is abstracted from red and nonred stimuli (Miller & Johnson-Laird, 1976). Although this may seem a trivial procedure, it is not; it requires complex information processing. It is difficult to program a computer to form abstract concepts. Observation of young children indicates that they are slower than expected to learn to name colors correctly; this would point to it being a complex and nontrivial process (Miller & Johnson-Laird, 1976).

For the second part, the concept is generalized or broadened beyond those sensations used in the abstraction process. Then, stimuli colored shades of red that have not been seen before, can be categorized as falling within the red concept; they have 'redness'. The concept is given a label, 'red', for the purposes of communication among those who share the concept. The same reasoning can be applied to the formation of concepts in the chemical or tactile senses: 'salty', 'sweet', 'fruity', 'crunchy', etc.

It can be argued that in everyday communication, because each person has his own life experience, sensory concepts would not correspond exactly between people. The argument whether a 'greeny-blue' colour should be categorized as 'green' or as 'blue' is not uncommon. People vary in how they distribute sensations among their concepts (Miller & Johnson-Laird, 1976; Ishii & O'Mahony, 1987a). Thus, without some sort of common training, it is to be expected that people would not communicate their sensations accurately enough for scientific research because they do not have their concepts aligned. Everyday language would not seem the medium for precise scientific communication of sensations; it is too vague. Certainly researchers have indicated the lack of commonality of everyday taste language (Peryam, 1957; Clapperton, 1975; Jenkins, 1980) pointing to the necessity for operational definition (Williams, 1975, 1981; Civille & Lawless, 1986).

What is needed is not *everyday* language but a *scientific* language. This is one in which every descriptive term is precisely defined. This means that all persons have the same set of concepts and the same labels for them. The Munsell Book of Color (1976) is equivalent to a scientific language. Here, each color is defined by a color plate in the book and labeled by a number. It is more like a dictionary than a spoken scientific language but the purpose is the same. It is worth examining the approaches adopted in sensory evaluation for establishing a scientific language.

DESCRIPTIVE ANALYSIS IN THE FOOD INDUSTRY

In sensory evaluation, it is sometimes desirable to be able to describe the sensory attributes of foods precisely. This is especially important in quality assurance, product development and assessment, correlation of instrumental and sensory properties, comparison of the sensory characteristics of products which cannot be compared simultaneously (e.g. fresh fruit from two consecutive years), studies of the effects of storage and packaging, etc. It follows that this is not the job for untrained judges because their sensory concepts would not be aligned. Thus, it would be unwise to use untrained consumers here because of their lack of training. It is a job for trained judges. Their training must include a procedure whereby their concepts are brought into line and each concept is given a common label for purposes of communication.

It is worth noting that taste psychophysicists often ignore this whole issue and allow judges to describe taste stimuli using their own idiosyncratic everyday language. Further, they only allow judges to use a restricted vocabulary of combinations of undefined descriptive terms: 'sweet', 'sour', 'salty', 'bitter', 'tasteless' and sometimes words like 'other'. This has been shown to cause lack of agreement among judges on their conceptual structures, as well as duplicate labelling of concepts (Ishii & O'Mahony, 1987b; O'Mahony & Ishii, 1987).

Food scientists have been more rational in their approach to concept alignment. There are several procedures for doing this. The earliest was the Flavor Profile Method, developed by the Arthur D. Little Company of Boston (Cairncross & Sjöström, 1950; Caul, 1957) while other well-known procedures are Quantitative Descriptive Analysis developed by the Tragon Corporation of Palo Alto, California (Stone et al., 1974) and the Spectrum Method developed by Sensory Spectrum Inc. of East Hanover, New Jersey (Meilgaard et al., 1987b). There are many variations on these techniques and it is not the intention of this chapter to describe precisely the procedures used; they are documented sufficiently in standard texts (Amerine et al., 1965; Gatchalian, 1981; Jellinek, 1985; Stone & Sidel, 1985; Meilgaard et al., 1987b).

In essence, these techniques create a common descriptive language using standards to define the descriptive labels. Judges examine the food under consideration and list all the relevant sensory characteristics. Then, they use standards to attain agreement on the descriptive terms being used. The procedure is an attempt to align the concepts as much as possible and agree on the labels for these concepts.

There are various scaling procedures used for estimating the intensity of the characteristic under consideration. Considerable time is spent in finding the right standards and learning which sensations fall into which concepts. Needless to say, the language developed is specific to the given product and the given panel of judges who underwent training. Generalization of the language to different products is seldom required and might not be desirable.

It is clear that this procedure aligns the centers of the concepts, the sensations from the standard stimuli around which the concepts are formed, but do they control the generalization step? How broad is the concept formed from a single standard stimulus? Do all judges generalize their concepts to the same degree? These are important questions because for clear communication about the sensory characteristics of a product, it is necessary to know the boundaries of the concepts so as to be able to determine whether a given sensation falls within a given concept or not. However, before being able to answer such questions, it is necessary to measure sensory concepts.

DESCRIPTIVE ANALYSIS: MEASUREMENT OF SENSORY CONCEPTS

A concept is an abstract entity. It is not like a neuron or a synapse. It can be thought of as a function of the software or information processing mechanisms of the brain. Although such an abstract entity cannot be isolated, the output generated by it can. A concept can be measured in terms of which stimuli elicit sensations that fall within it and which ones do not.

It is necessary to develop this argument further. The boundaries of a sensory concept do not appear to be clearly defined; they are fuzzy-edged. This means that although some stimuli elicit sensations that always fall within or outside the concept, there are other stimuli which elicit sensations that fall sometimes within the concept and sometimes outside. The color of blood may always fall within the red concept while the color black may always fall outside. Yet, a reddish-orange may sometimes be judged to fall within the concept and sometimes outside. Thus, if a concept were to be measured in terms of the probability of the sensation from a given stimulus falling within the concept, it should not be expected that the only probabilities obtained would be 100% and 0%. The concept can be visualized more as a cloud with fuzzy edges than as a balloon with sharp edges. There are even theories of fuzzy sets and rules for fuzzy inferences which can be used to model classification into fuzzy-edged concepts (Keuning et al., 1986).

So, the question becomes: How can a concept be measured? Given that it is fuzzy edged and that sensations from various stimuli will have various probabilities of falling within the concept, the procedure becomes straightforward. A judge merely categorizes a set of stimuli as falling within the concept or outside it; this will then give a measure of the concept in terms of its classifying output. However, the judge will be faced with the criterion question: How similar to the center of the concept must a stimulus be in order to be reported as being in the concept? The center of the concept might be defined by a standard stimulus presented beforehand or in everyday language it will be a remembered sensation. One approach to solving the criterion problem was used by Ishii & O'Mahony (1987a). They used the signal detection strategy of including sureness judgements in their question. Thus, a judge could respond: 'inside the concept, sure'; 'inside the concept, not sure'; 'outside the concept, not sure'; 'outside the concept, sure'. This gave a signal detection index for the degree of inclusion within (or exclusion from) the concept. The index used was the R index (Brown, 1974; O'Mahony, 1979) which measured the degree of inclusion within the concept in terms of probability. Thus, despite its abstract nature, a concept is susceptible to measurement.

Having established a method for measuring a concept, it was then possible for Ishii and O'Mahony to address some of the questions posed by the techniques of descriptive analysis. The question posed before was whether a single standard was sufficient to align the sensory concepts of all the judges. Would all judges generalize their concepts to the same extent? The answer was no. They measured the taste concepts formed from a single standard for each of twelve judges. The concepts were measured by assessing whether a set of fourteen stimuli fell within the concept so formed. The concept for each judge was expressed as a set of fourteen R indices, one for each test stimulus. It was found that all judges had generalized their concepts to different degrees; none had formed the same concept from a single standard.

It would seem logical, considering the normal concept formation mechanism, that judges might have their concepts aligned more if more standards were used to specify the boundaries of the concepts. Thus, judges could taste a set of standards to establish a concept, some falling within the concept and some falling outside. Ishii and O'Mahony

(1987a) made a beginning with this but the research is still in its infancy. Civille and Lawless (1986) came to the same conclusions about the desirability of multiple standards, independently.

Ishii and O'Mahony (1987a) also noted that the concepts formed tended to drift from day to day. This is interesting when one considers the day to day drift of EEG patterns measured in the rabbit's olfactory bulb, when it recognizes a particular odor (Skarda & Freeman, 1987). This is not to say that there is a direct correspondence between these two phenomena but they both indicate the lability of brain function and categorization processes.

The concept measurement technique can be applied in quality assurance. If a product can have a degree of variability in the marketplace, then forced choice difference tests in which the freshly made product is compared with a uniform standard would not be applicable. The freshly-made product might be different from the standard but still be acceptable as an example of that product by the consumer. Here, the consumer's concept of the product would be broader than a single standard. Instead of judges comparing the freshly made product with physical standards, they could compare them with conceptual standards. They could use Ishii and O'Mahony's (1987a) signal detection technique (R index) to determine the probability of the freshly-made product falling within the appropriate product concept. From this, a decision about whether to release the freshly prepared product on to the market could be made. Judges would need to realign their concept before each testing session by tasting standards, some of which fell within the concept and some outside. The appropriate standards could be determined from measures of consumer acceptability. At the time of writing, this approach has been adopted and is being used successfully by a European candy company.

With techniques available for measuring sensory concepts and the realization that descriptive analysis is based on concept formation, there are now methods available for assessing critically various descriptive analysis techniques. Concept measurement techniques also have the potential for use in other areas thus increasing the range of methods available to the sensory analyst. Study of concept formation in sensory evaluation would seem to be a worthwhile endeavour.

Sensory evaluation should not be a set of methods into which foods must be made to fit. A test procedure required for jalapeño peppers would not be expected to be the same as one for white bread or soap powder. It would seem better to design testing procedures especially for a given product. To do this, it is important to understand the physiological and cognitive limits of the judge. Thus, basic research into sensory function and the development of a wider array of methods based on this information is vital to the field of Sensory Evaluation.

REFERENCES

Amerine, M. A., Pangborn, R. M. & Roessler, E. B. (1965). *Principles of Sensory Evaluation of Food*. Academic Press, New York, pp. 321–48, 377–86.
Basker, D. (1980). Polygonal and polyhedral taste testing. *Journal of Food Quality*, **3**, 1–10.
Brown, J. (1974). Recognition assessed by rating and ranking. *British Journal of Psychology*, **65**, 13–22.
Buchanan, B., Givon, M. & Goldman, A. (1987). Measurement of discrimination ability in taste tests: An empirical investigation. *Journal of Marketing Research*, **24**, 154–63.
Byer, A. J. & Abrams, D. (1953). A comparison of the triangular and two sample taste test methods. *Food Technology*, **7**, 185–7.
Cairncross, S. E. & Sjöström, L. B. (1950). Flavor profiles — a new approach to flavor problems. *Food Technology*, **4**, 308–11.
Caul, J. (1957). The profile method in flavor analysis. *Advances in Food Research*, **7**, 1–40.
Civille, G. V. & Lawless, H. T. (1986). The importance of language in describing perceptions. *Journal of Sensory Studies*, **1**, 203–15.
Clapperton, J. F. (1975). The development of a flavour library. In *Proceedings of the 15th Congress, European Brewery Convention, Nice*. Elsevier, Amsterdam, pp. 832–5.
Coren, S., Porac, C. & Ward, L. M. (1979) *Sensation and Perception*. Academic Press, New York, pp. 17–23.
D'Amato, M. R. (1970). *Experimental Psychology*. McGraw-Hill, New York, pp. 157–84.
Dawson, E. H. & Dochterman, E. F. (1951). A comparison of sensory methods of measuring differences in food qualities. *Food Technology*, **5**, 79–81.
Dawson, E. H., Brogdon, J. L. & McManus, S. (1963). Sensory testing of differences in taste. 1. Methods. *Food Technology*, **17**, 1125–31.
Ennis, D. M. & Mullen, K. (1985). The effect of dimensionality on results from triangular method. *Chemical Senses*, **10**, 605–8.
Ennis, D. M. & Mullen, K. (1986a). Theoretical aspects of sensory discrimination. *Chemical Senses*, **11**, 513–22.
Ennis, D. M. & Mullen, K. (1986b). A multivariate model for discrimination methods. *Journal of Mathematical Psychology*, **30**, 206–19.
Ennis, D. M., Mullen, K. & Frijters, J. E. R. (1988). Variants of the method of triads: Unidimensional Thurstonian models. *British Journal of Mathematical and Statistical Psychology*, **41**, 25–36.

Fallis, N., Lasagna, L. & Tétreault, L. (1962). Gustatory thresholds in patients with hypertension. *Nature*, **196**, 74–5.

Filipello, F. (1956). A critical comparison of the two-sample triangular binomial designs. *Food Research*, **21**, 235–41.

Frijters, J. E. R. (1979). Variations of the triangular method and the relationship of its unidimensional probabilistic models to three-alternative forced-choice signal detection theory models. *British Journal of Mathematical and Statistical Psychology*, **32**, 229–41.

Frijters, J. E. R. (1981). An olfactory investigation of the compatibility of oddity instructions with the design of a 3–AFC signal detection task. *Acta Psychologica*, **49**, 1–16.

Frijters, J. E. R. (1984). Sensory difference testing and the measurement of sensory discriminability. In *Sensory Analysis of Foods*, ed. J. R. Piggott. Elsevier Applied Science Publishers, London, pp. 117–40.

Frijters, J. E. R., Blauw, Y. H. & Vermaat, S. H. (1982). Incidental training in the triangular method. *Chemical Senses*, **7**, 63–9.

Furchtgott, E. & Willingham, W. W. (1956). The effect of sleep-deprivation upon the threshold of taste. *American Journal of Psychology*, **69**, 111–12.

Gatchalian, M. M. (1981). *Sensory Evaluation Methods with Statistical Analysis*. College of Home Economics, University of the Philippines, Diliman, Quezon City, Philippines, pp. 143–77, 178–216.

Goldstein, E. B. (1984). *Sensation and Perception*. Wadsworth, Belmont, California, p. 12–18.

Green, D. M. & Swets, J. A. (1966). *Signal Detection Theory and Psychophysics*. John Wiley, New York.

Gridgeman, N. T. (1955). Taste comparisons: Two samples or three?, *Food Technology*, **9**, 148–50.

Gridgeman, N. T. (1956). Group size in taste sorting trials. *Food Research*, **21**, 534–9.

Grim, A. C. & Goldblith, S. A. (1965). Some observed discrepancies in application of the triangle test to evaluation of irradiated whole-egg magma. *Food Technology*, **19**, 146.

Harris, H. & Kalmus, H. (1949). The measurement of taste sensitivity to phenylthiourea (P.T.C.). *Annals of Eugenics*, **15**, 24–31.

Helm, E. & Trolle, B. (1946). Selection of a taste panel. *Wallerstein Laboratory Communications*, **9**, 181–94.

Hogue, D. V. & Briant, A. M. (1957). Determining flavor differences in crops treated with pesticides. I. A comparison of a triangle and multiple comparison method. *Food Research*, **22**, 351–7.

Hopkins, J. W. (1954). Some observations on sensitivity and repeatability of triad difference tests. *Biometrics*, **10**, 521–30.

Hopkins, J. W. & Gridgeman, N. J. (1955). Comparative sensitivity of pair and triad flavor intensity difference tests. *Biometrics*, **11**, 63–8.

Hull, C. L. (1920). Quantitative aspects of the evolution of concepts. *Psychological Monographs*, **28**, 1–86.

Ishii, R. & O'Mahony, M. (1987*a*). Defining a taste by a single standard: Aspects of salty and umami tastes. *Journal of Food Science*, **52**, 1405–9.

Ishii, R. & O'Mahony, M. (1987*b*). Taste sorting and naming: Can taste concepts be misrepresented by traditional psychophysical labelling systems. *Chemical Senses*, **12**, 37–51.
Jellinek, G. (1985). *Sensory Evaluation of Food*. Ellis Horwood, Chichester, pp. 184–251, 288–307.
Jenkins, L. A. (1980). Finding the common underlying continua from individual vocabularies. *Journal of the Science of Food and Agriculture*, **31**, 622.
Kapenga, J. A., de Doncker, E., Mullen, K. & Ennis, D. M. (1987). The integration of the multivariate normal density function for the triangular method. In *Numerical Integration*, ed. P. Keast & G. Fairweather. D. Reidel Publishing Co., Dordrecht, Holland, pp. 49–66.
Keuning, R., Backer, E., Duin, R. P. W., Lincklaen Westenberg, H. W. & de Jong, S. (1986). Fuzzy set theory applied to product classification by a sensory panel. *Chemical Senses*, **11**, 622–3.
Larmond, E. (1977) *Laboratory Methods for Sensory Evaluation of Food*. Canada Department of Agriculture, Publication 1637, Ottawa, Canada, pp. 20–37.
Matlin, M. W. (1988). *Sensation and Perception*. Allyn and Bacon, Boston, pp. 22–33.
McNicol, D. (1972). *A Primer of Signal Detection Theory*. George Allen and Unwin, London.
Meilgaard, M., Civille, G. V. & Carr, B. T. (1987*a*). *Sensory Evaluation Techniques*, Vol. I. CRC Press, Boca Raton, Florida, pp. 47–112.
Meilgaard, M., Civille, G. V. & Carr, B. T. (1987*b*). *Sensory Evaluation Techniques*, Vol. II. CRC Press, Boca Raton, Florida, pp. 1–23.
Miller, G. A. & Johnson-Laird, P. N. (1976). *Language and Perception*. Cambridge University Press, London.
Millward, R. B. (1980). Models of concept formation. In *Aptitude, Learning and Instruction, Volume 2, Cognitive Process Analyses of Learning and Problem Solving*, ed. R. E. Snow, P. A. Federico & W. E. Montague. Lawrence Erlbaum Associates, Hillsdale, New Jersey, pp. 245–75.
Mitchell, J. W. (1956). The effect of assignment of testing materials to the paired and odd position in the duo–trio taste difference test. *Food Technology*, **10**, 169–71.
Mullen, K., Ennis, D. M., de Doncker, E. & Kapenga, J. A. (1988). Models for duo–trio and triangular methods. *Biometrics*, **44**, 1169–75.
Munsell Book of Color (1976). MacBeth Division of Kollmorgen Corp, Baltimore, Maryland.
Nelson, K. (1974). Concept, word, and sentence. Interrelations in acquisition and development. *Psychological Review*, **81**, 267–85.
O'Mahony, M. (1972). Salt taste sensitivity: A signal detection approach. *Perception*, **1**, 459–64.
O'Mahony, M. (1979). Short-cut signal detection measures for sensory analysis. *Journal of Food Science*, **44**, 302–3.
O'Mahony, M. (1982). Some assumptions and difficulties with common statistics. *Food Technology*, **36**, 75–82.
O'Mahony, M. (1988). Sensory difference and preference testing: The use of signal detection measures. In *Applied Sensory Analysis of Foods*, ed. H. Moskowitz, CRC Press, Boca Raton, Florida, pp. 145–75.

O'Mahony, M. & Goldstein, L. R. (1986). Effectiveness of sensory difference tests: Sequential sensitivity analysis for liquid food stimuli. *Journal of Food Science*, **51**, 1550–3.

O'Mahony, M. & Goldstein, L. R. (1987a). Sensory techniques for measuring differences in California navel oranges treated with low doses of gamma-radiation below 0.6 kgray. *Journal of Food Sciences*, **52**, 348–52.

O'Mahony, M. & Goldstein, L. R. (1987b). Methods for sensory evaluation of navel oranges treated with electron beam irradiation. *Lebensmittel-Wissenschaft und Technologie*, **20**, 78–82.

O'Mahony, M. & Goldstein, L. R. (1987c). Tasting successive salt and water stimuli: The roles of adaptation, variability in physical signal strength, learning, supra- and subadapting signal detectability. *Chemical Senses*, **12**, 425–36.

O'Mahony, M. & Ishii, R. (1987). The umami taste concept: Implications for the dogma of four basic tastes. In *Umami: A Basic Taste*, ed. Y. Kawamura & M. R. Kare. Marcel Dekker, New York, 75–93.

O'Mahony, M. & Odbert, N. (1985). A comparison of sensory difference testing procedures: Sequential sensitivity analysis and aspects of taste adaptation. *Journal of Food Science*, **50**, 1055–8.

O'Mahony, M., Wong, S-Y. & Odbert, N. (1985a). Sensory evaluation of navel oranges treated with low doses of gamma-radiation. *Journal of Food Science*, **50**, 639–46.

O'Mahony, M., Wong, S-Y. & Odbert, N. (1985b). Initial sensory evaluation of bing cherries treated with low doses of gamma-radiation. *Journal of Food Science*, **50**, 1048–50.

O'Mahony, M., Wong, S-Y. & Odbert, N. (1985c). Sensory evaluation of regina freestone peaches treated with low doses of gamma radiation. *Journal of Food Science*, **50**, 1051–4.

O'Mahony, M., Wong, S-Y. & Odbert, N. (1986). Sensory difference tests: Some rethinking concerning the general rule that more sensitive tests use fewer stimuli. *Lebensmittel-Wissenschaft und Technologie*, **19**, 93–4.

Peryam, D. R. (1957). Factors affecting the accuracy and reliability of sensory tests. *American Society for Quality Control Transactions Annual Conventions*, pp. 675–85.

Peryam, D. R. (1958). Sensory difference tests. *Food Technology*, **12**, 231–6.

Peryam, D. R. & Swartz, V. W. (1950). Measurement of sensory differences. *Food Technology*, **4**, 390–5.

Pfaffmann, C., Schlosberg, H. & Cornsweet, J. (1954). Variables affecting difference tests. In *Food Acceptance Methodology*, Adv. Board Quartermaster Research and Development Committee on Foods, National Academy of Science National Research Council, Chicago, pp. 4–20.

Pikas, A. (1966). *Abstraction and Concept Formation*. Harvard University Press, Cambridge, MA.

Pokorný, J., Marcín, A. & Davídek, J. (1981). Comparison of the efficiency of triangle and tetrade tests for discrimination sensory analysis of food. *Die Nahrung*, **25**, 561–4.

Schelling, J. L., Tétreault, L., Lasagna, L. & Davis, M. (1965). Abnormal taste thresholds in diabetes. *Lancet*, **1**, 508–12.

Schiffman, H. R. (1982) *Sensation and Perception*. John Wiley, New York, pp. 7-14.

Skarda, C. A. & Freeman, W. J. (1987). How brains make chaos in order to make sense of the world. *Brain and Behavioral Sciences*, **10**, 161-95.

Spencer, H. W. (1979). Alternative sensory test designs. *Journal of the Science of Food and Agriculture*, **30**, 218-19.

Stone, H. & Sidel, J. L. (1985). *Sensory Evaluation Practices*. Academic Press, New York, pp. 132-93, 194-226.

Stone, H., Sidel, J., Oliver, S., Woolsey, A. & Singleton, R. C. (1974). Sensory evaluation by quantitative descriptive analysis. *Food Technology*, **28**, 24-34.

Topinka, I. & Sova, J. (1967). Taste threshold for NaCl, its genetic bond and possible significance in the etiology of hypertension. *Casopis Lékaru Ceských*, **106**, 689-93.

Wasserman, A. E. & Talley, F. (1969). A sample bias in the evaluation of smoked frankfurters by the triangle test. *Journal of Food Science*, **34**, 99-100.

Williams, A. A. (1975) The development of a vocabulary and profile assessment method for evaluating the flavor contribution of cider and perry aroma constituents. *Journal of the Science of Food and Agriculture*, **26**, 567-82.

Williams, A. A. (1981). Relating sensory aspects to quality. In *Quality in Stored and Processed Vegetables and Fruit*, ed. P. W. Goodenough & R. K. Atkins. Academic Press, London, pp. 17-33.

7

Attitudes and Beliefs as Determinants of Food Choice

R. SHEPHERD
*AFRC Institute of Food Research, Reading Laboratory,
Shinfield, Reading RG2 9AT, UK*

INTRODUCTION

Human food choice is determined by a large number of factors. Various schemes have been put forward which incorporate some of these influences (Pilgrim, 1957; Khan, 1981; Randall & Sanjur, 1981; Booth & Shepherd, 1988).

Shepherd (1985) used the scheme shown in Fig. 1. The factors are split into those related to the food, the person and the environment. The food has a particular chemical and physical composition, which give rise to properties of the food which are perceived by the individual as sensory characteristics, such as appearance, taste, odour and texture. Many aspects of the chemical composition of the food will not, however, be perceived. The sensory attributes of themselves do not lead to food acceptance or rejection, rather it will be the individual's liking or preference for particular foods or for the level of an attribute in a particular food, which will influence choice.

The chemical components of a food may also be thought of in terms of nutrients, such as protein, fat, carbohydrate, or minerals. Ingestion of these is important for the well being of the organism and therefore the chemical composition (and also the physical structure of the food) will have physiological effects following ingestion. These physiological effects will influence subsequent food selection (Booth *et al.*, 1982; Blundell *et al.*, 1987). Individual differences in other physiological factors related to the individual might also influence food selection, e.g. hormone levels, illness, intolerance to particular food constituents.

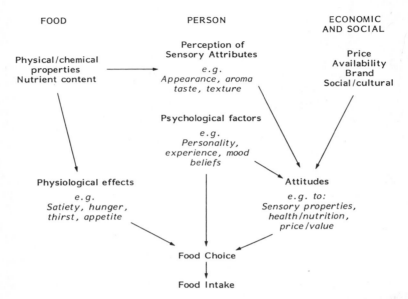

FIG. 1. A schematic diagram showing some factors influencing food choice (from Shepherd, 1985).

There will also be individual differences in psychological factors, such as personality, which may affect food choice (Jalso et al., 1965; Woolcott et al., 1983). Part of this effect might be through differences in the lifestyle of individuals with different personality types. Differences in levels of education and knowledge about nutrition, foods, preparation, etc., will lead to different food use. Differences between individuals in previous experience and learning associated with foods will lead to differences in beliefs, values and habits concerning particular foods. For particular occasions there will be foods which are considered more or less appropriate (Schutz et al., 1975; Schutz & Wahl, 1981). The post-ingestional effects of consuming certain types of foods might lead to psychological as well as physiological effects, e.g. mood, sleepiness; association between these effects and a particular food might influence subsequent selection of the food under appropriate circumstances.

External to both the individual and the food are the general social and cultural environment. Food choice varies between cultures (Katz, 1982; Simoons, 1982) and the social context surrounding the individual will also have an impact on food choice. Religion may require certain dietary choices to be made regardless of personal preferences. The availability, convenience of purchase, price, packaging, advertising and marketing will all have an influence.

Differences in age, sex, social class, region of residence, degree of urbanisation will all lead to differences in food consumption. These may operate through some of the other variables described above. Many of the variables will be interrelated and difficult to distinguish. Also food choice is not a constant phenomenon but will change with differing circumstances and with experiences of the individual.

ATTITUDES AND BELIEFS

Before looking more closely at the role played by beliefs in food acceptance, it is necessary to consider some form of definition of the terms used. Many of the problems in the literature on attitudes and beliefs stem from the lack of a clear and adequate definition. Allport (1935) defined an attitude as 'a mental and neural state of readiness to respond, organized through experience exerting a directive and/or dynamic influence on behavior'. Later workers (e.g. Krech & Crutchfield, 1948) have conceptualised attitudes in terms of three components: (a) cognitive, i.e. information or beliefs about the object; (b) affective, i.e. feelings of like or dislike towards the object; (c) conative, i.e. the tendency to behave in a certain way towards the object. Often the affective component is taken as the central component of attitude. Within this framework, beliefs may be thought of as the cognitive component. These beliefs may be seen either as a component of an overall attitude, or as something separate from, but closely related to, the affective attitude. It is, however, important that these different components are not confused as they have been in many studies.

Attitudes and food choice

The relative importance of influences on dietary choice might be assessed by asking subjects directly. Schafer (1978), for example, found that husbands rated taste as being the most important determinant of food choice with nutrition second; wives, on the other hand, placed nutrition as the most important determinant. McNutt *et al.* (1986) found subjects rated safety most important, followed by taste. One problem with this type of approach is that people may not be aware of the importance of different influences on their behaviour. Although in some instances people's expressed reasons for performing particular behaviours may be correct in other cases they may have little insight into why they perform them. It is therefore necessary to test the relationship between people's responses to particular foods and their consumption of them, rather than relying on their own perception of the importance of the different influences.

There have been a large number of studies of this type, which have been reviewed by Foley *et al.* (1979) and Khan (1981). The findings are not generally clear. There are two main approaches to investigating attitudes in this area. The first involves investigating preferences for the food or for specific sensory attributes of the food (Piggott, 1979; Randall & Sanjur, 1981), and has included much work on the food preferences of American soldiers (e.g. Peryam *et al.*, 1960). Work on the relationship between food choice and specific sensory attributes in foods has been reviewed by Shepherd (1988*b*).

In the nutritional literature the emphasis has been on attempts to relate nutritional knowledge, attitudes and behaviour. Attitudes and beliefs have generally been assessed in relation to nutrition rather than sensory or economic factors. It might be supposed that there will be a causal relationship between these variables, with a good nutritional knowledge leading to appropriate beliefs, positive attitudes towards nutritionally beneficial foods, and hence greater consumption of these foods. However, the strength of this relationship, assessed in studies in this area, has been somewhat weak.

A significant relationship between knowledge and attitudes has been found in several studies (Eppright *et al.*, 1970; Schwartz, 1975; Sims, 1976; Grotkowski & Sims, 1978; Werblow *et al.*, 1978; Foley *et al.*, 1983; Perron & Endres, 1985). In many of these studies general attitudes towards nutrition (e.g. nutrition is important) have been assessed. In some, the attitude component has apparently contained cognitive belief items, which would lead to a greater similarity between the knowledge and attitude items (both being cognitive) and hence inflate the observed relationship. A number of studies have used a measure of 'nutritional practices' as the behaviour of interest; this is generally a composite score from a number of questions such as whether the person eats breakfast or eats three proper meals per day, which together are taken as being positive nutritional practices. Where such measures are used, there tends to be a significant relationship between attitudes and behaviour (Jalso *et al.*, 1965; Schwartz, 1975; Carruth *et al.*, 1977; Foley *et al.*, 1983; Douglas & Douglas, 1984). If consumption of specific foods or intakes of specific nutrients are used as the appropriate measures of behaviour, either only a small number of statistically significant relationships are found (Eppright et al., 1970; Grotkowski & Sims, 1978), or there are none (Werblow et al., 1978; Perron & Endres, 1985). Knowledge and behaviour show an association when the behaviour is a general nutritional practice score (Carruth et al.,1977; Foley et al., 1983; Allen & Ries, 1985), but when the consumption of specific foods or nutrients is investigated there are again no significant relationships (Grotkowski & Sims, 1978; Perron & Endres, 1985) or relationships for only a few of the nutrients examined (Eppright et al., 1970).

It would thus appear that nutritional knowledge may be related to attitudes and to behaviour, when the behaviour is one of general dietary practice. This breaks down when the consumption of specific foods is assessed. Part of the problem in trying to understand these relationships is the confusion in the literature over the definitions of attitudes, beliefs, knowledge and the appropriate forms of behaviour to assess. It is necessary to have clear definitions for these and to have a framework within which to relate them. In a review of the literature on attitudes, Fishbein and Ajzen (1972) found some five hundred operational definitions. With such diversity it is difficult to develop any clear picture.

Krondl and Lau (1982) put forward an approach for studying the relative importance of different factors in influencing food choice. In a series of studies they investigated the perception by individuals of the price, convenience, prestige, health beliefs, and flavour of foods and how these relate to consumption. It is important to note that factors such as price are not actual price in this context but are the ratings given by subjects on a scale of 'very cheap' to 'very expensive'. Use is then made of the individual differences in perception of price for a particular food and a correlation is calculated for each food, with the expectation that those subjects saying that the price is high will be less likely to buy the food. These are therefore subjective indices rather than objective measures of price, etc.

Krondl, Lau and co-workers found that for the majority of foods examined the major determinant of consumption was the flavour (or taste) of the food. Physiological factors like tolerance and satiety were also often important but health beliefs tended to be less important, whilst price and convenience were found to be unimportant. Prestige was found to differ in importance between the different groups tested (Krondl & Lau, 1982).

This approach tends to include just a few simple questions on a large number of foods and to judge the importance by the number of significant correlations. Although there may be consistently different influences on choice for different types of foods, this work has not confirmed any such differences (Krondl & Lau, 1982). An alternative approach is to study the influences on the choice of particular foods in much more detail.

Recent work on food choice has included the use of attitude and belief models from social psychology, which offer a clear framework within which to examine the relationships between beliefs, attitudes and food choice. These approaches also have the advantage of being tested in a number of applications outside of the food area. One such model is that proposed by Fishbein and Ajzen (1975); this will be described and applications in the food choice area will then be discussed.

FISHBEIN AND AJZEN ATTITUDE MODEL

Fishbein and Ajzen (1975) presented a model incorporating measures of beliefs, attitude, behavioural intention and behaviour and a framework for relating them; this was later elaborated by Ajzen and Fishbein (1980). Within this model, beliefs and attitudes are assessed in relation to a particular behaviour rather than towards an object. In the case of food choice such a behaviour would be consuming, purchasing or eating particular foods or types of foods. The components of the Fishbein and Ajzen model are shown schematically in Fig. 2. These

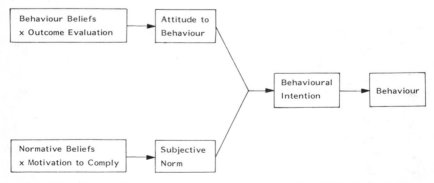

FIG. 2. The relationships between the components of the Fishbein and Ajzen attitudes model.

components, along with the relationships between them, will be explained in more detail below.

In this model, the best predictor of a person's behaviour is assumed to be behavioural intention; that is his or her conscious decision to perform that behaviour. Given that it relates to rational behaviour, it has been called the theory of reasoned action. Such an analysis will be less applicable where the behaviour is less under the conscious control of the individual. In different conditions it can be determined empirically how well intention predicts behaviour. It is assumed that there are choices open to the individual, e.g. choosing to buy one food rather than another; at the very least this would be a decision to perform the behaviour against the decision not to perform the behaviour. Behavioural intention is predicted by two components, the person's attitude towards the behaviour (A_{act}) and the subjective norm (SN). The attitude is an evaluation of whether the individual sees the behaviour as good or bad, harmful or beneficial, etc. The subjective norm is the perception by the individual of social pressure from important other people to perform the particular behaviour. The relationship between attitude, subjective norm and behavioural intention is summarised in the equation:

$$\text{Behavioural intention} = w_1 \times A_{act} + w_2 \times SN$$

The relative weightings (w_1 and w_2) for the components can be derived from the standardised (beta) coefficients from a multiple regression of the two components against behavioural intention. These weightings give an indication of the relative importance of the two components in predicting behavioural intention.

The attitude is in turn predicted by beliefs about the expected outcome of the behaviour. These are termed behaviour beliefs and for an individual there will be a relatively small set of such salient beliefs (b_i). These beliefs can be elicited from short interviews with a small group of people before a standard questionnaire is designed; it is important that these people should be similar to those to be tested with the standard questionnaire. The beliefs will not be the same for everyone, but the most frequently expressed beliefs can be used to derive a modal set, which will be included in the standard questionnaire.In addition to the beliefs about the outcome of the behaviour, the individual's evaluation of the expected outcome (e_i) will be important. Thus, an individual might say that eating a particular food will involve less preparation time, but if the individual is not at all concerned with preparation time, then this belief would be unlikely to exert much influence on this person's attitude or behaviour. Thus each belief score is multiplied by the appropriate evaluation score and these are then summed to give an overall belief-value measure. This relationship is expressed in the equation:

$$A_{act} = \sum b_i \times e_i$$

Because of the use of this type of function in the model, it is called an expectancy-value model. It makes use not only of beliefs about the expected outcome but also the value attached to such an outcome by the individual.

The subjective norm is predicted in a similar way from a set of normative beliefs (NB_j) about whether the individual thinks specific other people (e.g. spouse, doctor) or groups of people (e.g. government, medical profession) would prefer him or her to perform the behaviour. This therefore breaks down the subjective norm component to examine the roles played by specific others in influencing the individual. The normative beliefs are modified by how much the individual wants to comply with the perceived wishes of these other people. This is called the motivation to comply (Mc_j). Again there will be a set of salient beliefs for each individual but in order to generate a general set to be included in a standard questionnaire, a series of short interviews are used as for the behaviour beliefs. The normative beliefs multiplied by the motivation to comply are summed in order to predict the subjective norm:

$$SN = \sum NB_j \times Mc_j$$

This approach has been used widely in social psychology in relation to a large number of different forms of behaviour. In these various applications it has proved useful in showing the relationships between beliefs, attitudes and behaviour (e.g. Ajzen & Fishbein, 1980). It has recently been applied in the food choice area.

Application of the Fishbein and Ajzen model in food choice research

The Fishbein and Ajzen attitude model has been used in several studies of food selection. An early study by Bonfield (1974) showed relatively good prediction of the choice of soft drinks, with the attitudes and subjective norm components of the basic Fishbein and Ajzen model. Results were found to differ for different groups of subjects depending on such factors as education, income and the importance attached to product purchase.

Axelson et al. (1983) investigated eating in fast-food restaurants. They found behavioural intention was well predicted by the attitude component but the subjective norm component did not significantly add to this prediction. The summed belief-evaluation score predicted the attitude component.

Tuorila-Ollikainen et al. (1986) studied choice of low-salt bread, calculating the regression of the belief items directly against behavioural intention. The belief items related to the sensory attributes of the bread. The belief-value items were more closely related to behavioural intention than were the normative beliefs. The study involved an experimental period during which the subjects also tasted bread samples and gave hedonic ratings and made choices of which types of bread to consume. The relationships were similar both at the beginning and end of this period. Inclusion of the hedonic ratings into the Fishbein and Ajzen model increased prediction both of intention and of choice behaviour. Thus despite the Fishbein and Ajzen model covering the liking for sensory attributes of the bread, tasting the samples gave extra information independent of these ratings in determining intention and choice. If such a finding were general for all food choices, it would mean that the Fishbein and Ajzen model was not adequately assessing the liking component for foods and that this liking for the sensory aspects of the foods was exerting an independent influence on both intention and choice.

Tuorila (1987) investigated consumption of milks varying in fat content. Beliefs were included, which could be divided into general groups of sensory, nutritional, suitability and price items. Except for price, these all correlated highly with the attitude score. Likewise the beliefs were most positive for the type of milk generally consumed, indicating the role played by these beliefs in determining choice. The normative beliefs for family members related to the subjective norm score more than did those for other groups. Hedonic ratings of milk samples were not found to improve the prediction of intention over the basic Fishbein and Ajzen model.

These two studies provide conflicting evidence on whether hedonic responses to foods actually tasted give information not present in the Fishbein and Ajzen model. It might be that tasting the samples gives extra information, independent of the Fishbein assessed attitude, only for a food with which the subjects are not especially familiar, but for familiar foods the attitude score reflects the liking, and tasting actual samples does not add to this general attitude.

Feldman and Mayhew (1984) studied meat and sodium consumption using a model derived from that of Fishbein and Ajzen, but incorporating aspects of the Health Beliefs Model (Becker et al., 1974) and Triandis (1977) model. In addition to intention, measures of habit and facilitating conditions (incorporating knowledge about the behaviour, ability to perform the behaviour and arousal) were included in the equation to predict self-reported behaviour. Each component made a contribution to this prediction but facilitating conditions were less important than either intent or habit. Behavioural intention was predicted reasonably well by a variation of the Fishbein and Ajzen model, in which attitude, belief-evaluations, normative belief-motivations and a measure of personal norms were entered separately into the equation. Lewis and Booth (1985) likewise incorporated elements of the Fishbein and Ajzen model with aspects of the Health Beliefs Model to study dieting and weight loss.

Schifter and Ajzen (1985) have used the theory of planned behaviour also to study weight loss. This approach includes a measure of perceived control, similar to the facilitating conditions included by Feldman and Mayhew (1984). Perceived control significantly improved prediction of intention from the attitude and subjective norm. Perceived control also added to prediction of weight loss from behavioural intention, implying that where subjects do not have total control their perception of the degree of control is important independently of intention. In this study the criterion variable was weight loss which is not really a behaviour but rather the result of certain behaviours. It remains to be seen whether perceived control is also important in predicting other forms of more clearly defined behaviours.

Shepherd and Stockley (1985) reported a study on the consumption of foods contributing highly to fat intake in the average UK diet. Attitude was found to be a better predictor of behaviour than the subjective norm component. The questionnaire measure of frequency of consumption was also reasonably predictive of measured intake of fat as a percentage of energy over a 7-day period in a separate group of 30 subjects ($r = 0.58$). These findings were replicated by Shepherd and Stockley (1987). They failed to show any relationship between nutritional knowledge and either attitude or behaviour. This may have been because the nutritional knowledge questionnaire was not sufficiently detailed to give an adequate indication of each individual's knowledge. Alternatively it might be that people do not make use of their knowledge in forming attitudes or in deciding on their behaviour. Further examination of this relationship within this framework would be warranted.

The above studies on fat did not include belief items. A further study of consumption of savoury snack foods included belief items in addition to assessing the attitudes and subjective norm components. The questionnaire was completed by 118 subjects, and again the attitude to behaviour was more important than the subjective norm, which was not statistically significant. The multiple correlation predicting behaviour was 0·53, which was hardly increased over the simple correlation with the attitude score alone ($r = 0·52$). Belief-evaluation predicted the attitude score with a correlation of $r = 0·59$, and the normative beliefs × motivation to comply predicted the subjective norm score ($r = 0·45$). Thus the components of the model were related in the manner expected, with the attitude being the more important predictor of behaviour.

Table salt use

Shepherd and Farleigh (1986a) investigated attitudes towards adding table salt to foods and Shepherd and Farleigh (1986b) tested further subjects with the same questionnaire. The data from all 117 subjects are reported here.

In order to generate belief items, short interviews were carried out prior to these studies, using people not included in either study. Five belief items were chosen, which accounted for 81% of the interview responses. The belief items were that adding table salt to food (a) makes it taste better, (b) increases the risk of ill health, (c) replaces lost body salt, (d) increases the risk of high blood pressure, (e) improves the food (other than taste). Seven-category response scales labelled 'Disagree' and 'Agree' at the extremes were used, with five corresponding outcome evaluations labelled 'Good' and 'Bad' at the extremes.

Five normative belief items were chosen from the same interviews, in this case accounting for 90% of the responses. These normative beliefs related to (a) members of the family (other than parents), (b) doctors, (c) food manufacturers, (d) parents, (e) nutritionists and dietitians. There were five corresponding items on motivation to comply. The questionnaire also included three attitude questions on whether adding table salt to food was good/bad, unpleasant/pleasant and harmful/beneficial. The subjective norm item was on whether most peope who

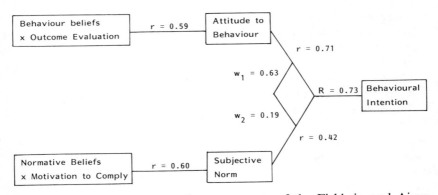

FIG. 3. Relationships between the components of the Fishbein and Ajzen model for table salt use ($n = 117$).

are important to the individual think he or she should add table salt to food. One question on the frequency of adding table salt to food served as a questionnaire measure of behaviour.

There was good prediction of table salt use from the components of the model, with correlations between components shown in Fig. 3. Thirty-six subjects took part in the study where table salt use was measured over 7 days using pre-weighed salt pots. This usage was well predicted by the questionnaire report of behaviour ($r = 0.64$).

Low-fat milk

In the basic Fishbein and Ajzen model it is assumed that the beliefs form a consistent pattern, i.e. if a person holds one negative belief (relative to other people) about a behaviour then the other beliefs will also be negative. This type of internal consistency is often assessed using measures such as Cronbach's alpha coefficient (e.g. McKennell, 1970). However, with choice of foods such a unitary belief structure might not be expected. An individual could consider that consuming a particular food gives sensory pleasure whilst still believing that the food is bad for health. This can be examined by factor analysing the belief responses to determine whether they are all consistent or whether they form separable factors.

Axelson *et al.* (1983) performed a factor analysis of the belief items in a study of eating in fast-food restaurants. They found two separable factors, described as organoleptic-nutritious and economic. There were some beliefs such as having limited selection, involving no cooking and the location of the restaurant, which did not load highly on either factor. There were thus separate factors in the beliefs in this study.

The belief structure related to consumption of low-fat milk was investigated by Shepherd (1988a) using a modification of the questionnaire developed by Tuorila (1987). One hundred and three subjects completed this questionnaire. It included questions on the amount of skimmed milk, semi-skimmed milk and whole-fat (silver top) milk consumed in an average week. There were three corresponding questions on behavioural intention to purchase each type of milk when next shopping for milk. Four attitude items were included on whether buying low-fat milk was bad/good, unpleasant/pleasant, harmful/beneficial and undesirable/desirable.

Tuorila (1987) found that the subjective norm was not important for this type of behaviour and hence the subjective norm and normative belief items were excluded. There were thirteen items on beliefs about the behaviour, which could be broken down into four groups: sensory attributes, nutritional, functional properties and price/value (see Table 1). The responses were on 7-category scales labelled 'Strongly disagree' and 'Strongly agree' at the extremes, with corresponding outcome evaluation items labelled 'Bad' and 'Good' at the extremes. All responses were scored from -3 to $+3$, with a positive score corresponding to a positive feeling towards consuming low-fat milk.

The relative intention towards low-fat milks was scored using the sum of the scores for skimmed and semi-skimmed milks minus that for whole-fat milk. The sum of all thirteen belief-evaluation scores correlated with the attitude score at $r = 0.80$ as shown in Fig. 4. The attitude score was correlated with behavioural intention ($r = 0.70$) and this was related to the score for behaviour ($r = 0.68$).

A principal components analysis of the belief-evaluation scores was used to test how well the belief items did fall into the supposed separate groups. The two component solution accounted for 52% of the variance, with the loadings shown in Table 1. The solution did not show evidence for the four original groups. Instead the nutritional items formed a

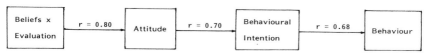

FIG. 4. Relationships between the components of the model for low-fat milk consumption ($n = 103$).

Table 1
Component loadings from principal components analysis of belief-evaluations from the low-fat milk questionnaire ($n = 103$). All responses were scored to have a high score for a positive belief-evaluation.

Questionnaire item	Component 1	Component 2
Sensory		
Tastes nice	0·70	0·40
Tastes watery	0·67	0·18
Can have off flavours	0·50	0·11
Leaves a greasy taste in the mouth	0·55	0·00
Nutritional		
Healthy	0·21	0·84
Fattening	0·12	0·77
Reduces my fat consumption	0·06	0·75
Functional		
Suitable for drinking by itself	0·63	0·35
Suitable for cooking	0·62	0·10
Suitable for tea or coffee	0·73	0·23
Price		
Expensive	0·04	0·21
Stays fresh for a long time	0·21	−0·06
Good value for money	0·52	0·28

separate component (component 2) from the sensory and functional items (component 1). The price items were not clearly loaded on either component and even in the three component solution where the items on 'stays fresh' and 'good value for money' loaded on the third component, the item on being expensive was not related to any of the components. This probably reflects the lack of a price differential for milks of different fat content.

Scores were calculated for two components (referred to as sensory/functional and nutritional) by summing the belief-evaluation scores for those items with loadings of 0·5 or greater on each of the components (see Table 1). A multiple regression of the sensory/functional and nutritional scores against attitude was used to test the relative importance of these components in predicting attitude. The multiple correlation of $R = 0.83$ showed good prediction of the attitude score.

The beta coefficients from the multiple regression were 0·44 for the sensory/functional and 0·53 for the nutritional beliefs. The simple correlations were 0·69 and 0·73 respectively. The nutritional component was thus more closely related to the attitude.

These analyses show that the beliefs are not unidimensional but tend to differentiate between nutritional and other factors. The nutritional beliefs tended to relate more closely to the attitude towards the choice of low-fat milks. This implies that people choosing low-fat milks do so for health reasons more so than because of the sensory/functional attributes of such milks, although these are still important. This would be likely to vary between foods and such variation needs to be investigated further.

Modifications of the basic approach
There are a number of problems associated with applying the basic model to the food choice area and hence modifications of the basic model might prove useful. Some of these would relate to any form of behaviour but others are specific to food choice.

As was argued earlier, there are a large number of factors which will influence food choice. Ajzen and Fishbein (1980) argue that these will act only through differences in beliefs, attitudes or subjective norms rather than contributing independently to the prediction of intention or behaviour. A number of studies have looked at this, generally finding it to be true (e.g. Schlegel *et al.*, 1977; Ajzen & Fishbein, 1980; Budd *et al.*, 1984). However, where the factors might not act independently of the components of this model, it is important to test how factors such as personality, age, socioeconomic class might operate through their effects on the components of the model. This should give a clearer understanding of the role played by these variables. Within this approach differences in the importance of factors between groups of subjects can be investigated.

There is some controversy about exactly how the components of the model are in fact related. For example, Grube *et al.* (1986) suggested that interactions between attitude and the subjective norm may need to be taken into account in addition to their purely additive effect. Bentler and Speckart (1979) and Katz (1985) suggested that attitude might exert an influence on behaviour which is not mediated by behavioural intention. Such modifications need to be examined.

There are a number of other models of social behaviour, such as that put forward by Triandis (1977). This includes a measure of habit as a predictor of behaviour along with intention and facilitating conditions. Intention is predicted by attitude, perceived value of consequences and social determinants. Whilst it is similar to the Fishbein and Ajzen model, some of its components are conceptualised in a different way and the equations relating the components are different. Brinberg (1979) tested both models and found both to give similar degrees of prediction of church attendance.

Bentler and Speckart (1979) suggested that past behaviour is an independent predictor of behavioural intention along with attitude and subjective norm, and also an independent predictor of behaviour in addition to intention. This would be similar to the concept of habit in the Triandis model. These predictions were confirmed in studies of some applications, such as drug and alcohol use (Bentler & Speckart, 1979) and seat belt use (Budd *et al.*, 1984). This would be worth investigating with behaviours like the buying or eating of foods which are performed very frequently. Here it might be that past behaviour is the important thing and that the individual does not have strong beliefs and attitudes but rather rationalises based on past behaviour. It would therefore be better to incorporate some measure of past behaviour or habit.

As was discussed above, Shifter and Ajzen have applied the theory of planned action to studying weight loss. There have been a number of other such applications outside of the area of food selection (e.g. Ajzen & Madden, 1986). The incorporation of measures of perceived control is similar to the concept of facilitating conditions in the Triandis model and it might usefully be incorporated into studying food choice.

Warshaw & Davis (1985) have suggested that intention is less appropriate as a predictor of behaviour than is expectation, i.e. people expect they will act in a particular way rather than deciding they will intentionally act in that way. Expectation might incorporate a number of the concepts such as facilitating conditions or perceived control which might intervene between intention and behaviour. Again this concept might usefully be investigated within the food choice area.

The basic model assumes that the belief structure is unidimensional. However, for some behaviours individuals might hold both positive and negative beliefs about different aspects of the behaviour. This may be particularly important for food choice where the factors influencing such choice might be rather disparate and not form a coherent and consistent belief structure. Grube *et al.* (1986) investigated smoking and found that in this case the belief structure was not unidimensional.

One problem with applying general models of choice to foods is the area of hedonics and pleasure associated with eating. The Fishbein and Ajzen model does not sufficiently capture the pleasurable aspects of eating and fails to take account of the liking for sensory attributes of particular samples of a food. In the food area the interest is often in samples of a food with slight variations in sensory or other attributes. In this case people's expectations about these would not differ until they have actually tested the samples. Including preference and hedonic assessment of samples of the food along with more general attitudinal measures might be more useful than either alone. Some progress has been made in this area. Tuorila-Ollikainen et al. (1986) found hedonic assessments of low-salt bread gave information additional to the belief-evaluations in predicting intention. However, this was not found in a study of milks with varying fat levels by Tuorila (1987). This might be because of the difference in the experience that the subjects have had with the type of food and hence the accuracy of their expectations. In the situation of purchasing and eating foods in the real world it might be that the expectations have more influence than hedonic assessments, since people with negative expectations will not try the product and hence not be able to compare the sensory attributes of that product with an alternative product. The interplay between attitudes and expectations and the sensory responses to actual foods (including hedonic responses) therefore requires much closer attention in the future.

Finally any attempt to understand the role played by beliefs and attitudes in food selection must offer guidance on how such beliefs, attitudes and behaviour might be changed. Such changes might be brought about by education or by advertising or any other means for changing attitudes and beliefs (McGuire, 1969). The Fishbein and Ajzen model gives indications of which beliefs are important in relation to a behaviour and hence which would be most reasonable to attempt to change. There have been successful attempts to change beliefs, attitudes and behaviour within this framework (e.g. Fishbein *et al.*, 1980), although not all such attempts have shown effects better than alternative procedures (e.g. Hoogstraten *et al.*, 1985). The usefulness of this approach in identifying salient beliefs which can then be changed through education and information needs much more empirical work. This method does offer both a means for identifying the beliefs and for assessing any changes in beliefs, attitudes and behaviour.

CONCLUSIONS

Food choice is affected by a large number of factors. These can be investigated in isolation but few would argue that any single influence will be all important. Whilst there have been a number of attempts to identify the different influences on food choice, these tend to be only descriptive and offer no means for investigating the different influences simultaneously. It can be argued that a number of the influences will act through beliefs and attitudes held by the individual but many of the attempts to relate beliefs, attitudes and behaviour have lacked a framework within which to examine these relationships adequately.

The attitudes model presented by Fishbein and Ajzen is one such framework and it has been successfully used in a number of applications in the food choice area. It offers clear operationalisation of the different components of the model and how these would be expected to relate to each other. It has shown good prediction of behaviour from behavioural intention. In general, people's own attitudes tend to outweigh perceived social pressure to behave in a particular way. The belief-evaluations relate to attitudes but there is some suggestion that the beliefs are not unidimensional and might have different components. This approach offers a means for assessing the relative importance of the different types of influence on food choice and this is an area where more can be done.

Although the model has proved useful in investigations in this area, there have been suggestions of modifications of the basic model in the attitude literature and these need to be explored in relation to food choice. Likewise, there are specific factors in food choice which might require extensions of the basic approach making it more food specific.

REFERENCES

Ajzen, I. & Fishbein, M. (1980). *Understanding Attitudes and Predicting Social Behavior*. Prentice-Hall, Englewood Cliffs, N.J.

Ajzen, I. & Madden, T. J. (1986). Prediction of goal-directed behavior: Attitudes, intentions, and perceived behavioral control. *J. Exp. Soc. Psychol.*, **22**, 453–74.

Allen, C. D. & Ries, C. P. (1985). Smoking, alcohol, and dietary practices during pregnancy: Comparison before and after prenatal education. *J. Am. Diet. Assoc.*, **85**, 605–6.

Allport, G. W. (1935). Attitudes. In *A Handbook of Social Psychology*, ed. C. Murchison. Clark University Press, Worcester, MA, pp. 789–844.
Axelson, M. L., Brinberg, D. & Durand, J. H. (1983). Eating at a fast-food restaurant—A social-psychological analysis. *J. Nutr. Educ.*, **15**, 94–8.
Becker, M. H., Drachman, R. H. & Kirscht, J. P. (1974). A new approach to explaining sick-role behavior in low income populations. *Am. J. Public Health*, **64**, 205–16.
Bentler, P. M. & Speckart, G. (1979). Models of attitude–behavior relations. *Psychol. Rev.*, **86**, 452–64.
Blundell, J. E., Rogers, P. J. & Hill, A. J. (1987). Evaluating the satiating power of foods: Implications for acceptance and consumption. In *Food Acceptance and Nutrition*, ed. J. Solms, D. A. Booth, R. M. Pangborn & O. Raunhardt. Academic Press, London, pp. 205–19.
Bonfield, E. H. (1974). Attitude, social influence, personal norms, and intention interactions as related to brand purchase behavior. *J. Marketing Res.*, **11**, 379–89.
Booth, D. A. & Shepherd, R. (1988). Sensory influences on food acceptance: The neglected approach to nutrition promotion. *BNF Nutr. Bull.*, **13**, 39–54.
Booth, D. A., Mather, P. & Fuller, J. (1982). Starch content of ordinary foods associatively conditions human appetite and satiation, indexed by intake and eating pleasantness of starch-paired flavours. *Appetite*, **3**, 163–84.
Brinberg, D. (1979). An examination of the determinants of intention and behavior: A comparison of two models. *J. Appl. Soc. Psychol.*, **9**, 560–75.
Budd, R. J., North, D. & Spencer, C. (1984) Understanding seat-belt use: A test of Bentler and Speckart's extension of the 'theory of reasoned action'. *Europ. J. Soc. Psychol.*, **14**, 69–77.
Carruth, B. R., Mangel, M. & Anderson, H. L. (1977). Assessing change-proneness and nutrition-related behaviors. *J. Am. Diet. Assoc.*, **70**, 47–53.
Douglas, P. D. & Douglas, J. G. (1984) Nutrition knowledge and food practices of high school athletes. *J. Am. Diet. Assoc.*, **84**, 1198–202.
Eppright, E. S., Fox, H. M., Fryer, B. A., Lamkin, G. H. & Vivian, V. M. (1970). The North Central Regional Study of diets of preschool children. 2. Nutrition knowledge and attitudes of mothers. *J. Home Econ.*, **62**, 327–32.
Feldman, R. H. L. & Mayhew, P. C. (1984). Predicting nutrition behavior: The utilization of a social psychological model of health behavior. *Basic Appl. Soc. Psychol.*, **5**, 183–95.
Fishbein, M. & Ajzen, I. (1972). Attitudes and opinions. *Ann. Rev. Psychol.*, **23**, 487–544.
Fishbein, M. & Ajzen, I. (1975). *Belief, Attitude, Intention and Behavior: An Introduction to Theory and Research*. Addison-Wesley, Reading, MA.
Fishbein, M., Ajzen, I. & McArdle, J. (1980). Changing the behavior of alcoholics: Effects of persuasive communication. In *Understanding Attitudes and Predicting Social Behavior*, ed. I. Ajzen & M. Fishbein. Prentice-Hall, Englewood Cliffs, N.J., pp. 217–42.
Foley, C., Hertzler, A. A. & Anderson, H. L. (1979). Attitudes and food habits—A review. *J. Am. Diet. Assoc.*, **75**, 13–18.
Foley, C. S., Vaden, A. G., Newell, G. K. & Dayton, A. D. (1983). Establishing the

need for nutrition education: III. Elementary students' nutrition knowledge, attitudes, and practices. *J. Am. Diet. Assoc.*, **83**, 564–8.

Grotkowski, M. L. & Sims, L. S. (1978). Nutritional knowledge, attitudes and dietary practices in the elderly. *J. Am. Diet. Assoc.*, **72**, 499–506.

Grube, J. W., Morgan, M. & McGree, S. T. (1986). Attitudes and normative beliefs as predictors of smoking intentions and behaviours: A test of three models. *Brit. J. Soc. Psychol.*, **25**, 81–93.

Hoogstraten, J., Haan, W. de and Horst, G. ter. (1985). Stimulating the demand for dental care: An application of Ajzen and Fishbein's theory of reasoned action. *Europ. J. Soc. Psychol.*, **15**, 401–14.

Jalso, S. B., Burns, M. M. & Rivers, J. M. (1965). Nutritional beliefs and practices. *J. Am. Diet. Assoc.*, **47**, 263–8.

Katz, J. (1985). The role of behavioral intentions in the prediction of behavior. *J. Soc. Psychol.*, **125**, 149–55.

Katz, S. H. (1982). Food, behavior and biocultural evolution. In *The Psychobiology of Human Food Selection*, ed. L. M. Barker. Ellis Horwood, Chichester, pp. 171–88.

Khan, M. A. (1981). Evaluation of food selection patterns and preferences. *CRC Crit. Rev. Food Sci. Nutr.*, **15**, 129–53.

Krech, D. & Crutchfield, R. S. (1948). *Theory and Problems in Social Psychology*. McGraw-Hill, New York.

Krondl, M. & Lau, D. (1982). Social determinants in human food selection. In *The Psychobiology of Human Food Selection*, ed. L. M. Barker. Ellis Horwood, Chichester, pp. 139–51.

Lewis, V. J. & Booth, D. A. (1985). Causal influences within an individual's dieting thoughts, feelings and behaviour. In *Measurement and Determinants of Food Habits and Food Preferences*, ed. J. M. Diehl & C. Leitzmann. EURO-NUT Report 7, Wageningen, The Netherlands, pp. 187–208.

McGuire, W. J. (1969). The nature of attitudes and attitude change. In *The Handbook of Social Psychology*, 2nd Edition, Vol. 3, ed. G. Lindzey & E. Aronson. Addison-Wesley, Reading, MA, pp. 136–314.

McKennell, A. (1970). Attitude measurement: Use of coefficient alpha with cluster or factor analysis. *Sociology*, **4**, 227–45.

McNutt, K. W., Powers, M. E. & Sloan, A. E. (1986). Food colors, flavors, and safety: A consumer viewpoint. *Food Technol.*, **40**, 72–8.

Perron, M. & Endres, J. (1985). Knowledge, attitudes and dietary practices of female athletes. *J. Am. Diet. Assoc.*, **85**, 573–6.

Peryam, D. R., Polemis, B. W., Kamen, J. M., Eindhoven, J. & Pilgrim, F. J. (1960). *Food Preferences of Men in the US Armed Forces*. Quartermaster Food and Container Institute, Chicago, Illinois.

Piggott, J. R. (1979). Food preferences of some United Kingdom residents. *J. Hum. Nutr.*, **33**, 197–205.

Pilgrim, F. J. (1957). The components of food acceptance and their measurement. *Am. J. Clin. Nutr.*, **5**, 171–5.

Randall, E. & Sanjur, D. (1981). Food preferences – Their conceptualization and relationship to consumption. *Ecol. Food Nutr.*, **11**, 151–61.

Schafer, R. B. (1978). Factors affecting food behavior and the quality of husbands' and wives' diets. *J. Am. Diet. Assoc.*, **72**, 138–43.

Schifter, D. E. & Ajzen, I. (1985). Intention, perceived control, and weight loss: An application of the theory of planned behavior. *J. Pers. Soc. Psychol.*, **49**, 843–51.

Schlegel, R. P., Crawford, C. A. & Sanborn, M. D. (1977). Correspondence and mediational properties of the Fishbein model: An application to adolescent alcohol use. *J. Exp. Soc. Psychol.*, **13**, 421–30.

Schutz, H. G. & Wahl, O. L. (1981). Consumer perception of the relative importance of appearance, flavor and texture to food acceptance. In *Criteria of Food Acceptance. How Man Chooses What He Eats*, ed. J. Solms & R. L. Hall. Forster, Zurich, pp. 97–116.

Schutz, H. G., Rucker, M. H. & Russell, G. F. (1975). Food and food-use classification system. *Food Technol.*, March, 50–64.

Schwartz, N. E. (1975). Nutritional knowledge, attitudes, and practices of high school graduates. *J. Am. Diet. Assoc.*, **66**, 28–31.

Shepherd, R. (1985). Dietary salt intake. *Nutr. Food Sci.*, **96**, (Sept/Oct), 10–1.

Shepherd, R. (1988*a*). Belief structure in relation to low-fat milk consumption. *J. Hum. Nutr. Diet.*, **1**, 421–8.

Shepherd, R. (1988*b*). Sensory influences on salt, sugar and fat intake. *Nutr. Res. Rev.*, **1**, 125–44.

Shepherd, R. & Farleigh, C. A. (1986*a*). Attitudes and personality related to salt intake. *Appetite*, **7**, 343–54.

Shepherd, R. & Farleigh, C. A. (1986*b*). Preferences, attitudes and personality as determinants of salt intake. *Hum. Nutr.: Appl. Nutr.*, **40A**, 195–208.

Shepherd, R. & Stockley, L. (1985). Fat consumption and attitudes towards food with a high fat content. *Hum. Nutr.: Appl. Nutr.*, **39A**, 431–42.

Shepherd, R. & Stockley, L. (1987). Nutrition knowledge, attitudes, and fat consumption. *J. Am. Diet. Assoc.*, **87**, 615–19.

Simoons, F. J. (1982). Geography and genetics as factors in the psychobiology of human food selection. In *The Psychobiology of Human Food Selection*, ed. L. M. Barker. Ellis Horwood, Chichester, pp. 205–24.

Sims, L. S. (1976). Demographic and attitudinal correlates of nutrition knowledge. *J. Nutr. Educ.*, **8**, 122–5.

Triandis, H. C. (1977). *Interpersonal Behavior*. Brooks/Cole, Monterey, CA.

Tuorila, H. (1987). Selection of milks with varying fat contents and related overall liking, attitudes, norms and intentions. *Appetite*, **8**, 1–14.

Tuorila-Ollikainen, H., Lahteenmaki, L. & Salovaara, H. (1986). Attitudes, norms, intentions and hedonic responses in the selection of low salt bread in a longitudinal choice experiment. *Appetite*, **7**, 127–39.

Warshaw, P. R. & Davis, F. D. (1985). Disentangling behavioral intention and behavioral expectation. *J. Exp. Soc. Psychol.*, **21**, 213–28.

Werblow, J. A., Fox, H. M. & Henneman, A. (1978). Nutritional knowledge, attitudes, and food patterns of women athletes. *J. Am. Diet. Assoc.*, **73**, 242–5.

Woolcott, D. M., Sabry, J. H. & Kawash, G. F. (1983). A study of some aspects of food-related behaviour among a group of men. *Hum. Nutr.: Appl. Nutr.*, **37A**, 199–209.

8

Designing Products for Individual Customers

D. A. BOOTH
School of Psychology, University of Birmingham, Edgbaston, Birmingham B15 2TT, UK

INTRODUCTION

The practical aim of sensory evaluation is to design products so that they appeal to potential purchasers. Psychology is the fundamental and applied science concerning the causal processes that operate in a person who, for example, is choosing among the range of available brands. Hence, it should be possible to find or create cost-effective psychological methods that sensory evaluators can use to guide competitively successful production and marketing of goods.

Such research guidance must relate factors that the producer, manufacturer and marketer can control to what actually goes on in the minds of individual customers buying and using the brand under investigation.

PERSONALISED DESIGN INFORMATION

Thus, the title of this chapter refers to research methods that personalise the information that goes into the design of goods. The goods themselves may well be mass-produced and mass-marketed, rather than customised. Advocacy of an individualised approach to sensory evaluation and consumer research has no particular affinity with personalisation of the item sold to the customer. The point is that a standard item still has to serve many individual customers' different needs.

The traditional shopkeeper prided himself on finding something to satisfy each customer. Such individual trading still occurs, with clothing for example. Also, customers can sometimes 'mix to taste', from bins of foodstuff variants, for example, or with features of automobile models. Up to now, however, these choices have always been very limited. Genuine personalisation (beyond the mere addition of the owner's name) may yet come, though, from advances in the technology of manufacturing and communications. Enough customers would have to be willing and able to pay for design, production and distribution to their own preferences. An ambition to design and sell items suited to individual tastes may then not be misplaced.

Nevertheless, individualised product design (or indeed mixing to taste) is relevant here not as a marketing strategy, but as a business intelligence technique. We find out what consumers want by finding out, within an adequately representative sample, what each individual customer wants.

Physical and verbal specification

On occasion, 'mix to taste' (the psychophysicists' production method) does happen to be a practicable research technique. The customer can prepare a physical mock-up or at least a pictorial sketch of the preferred form of the product.

More often, it is practicable only for the customer to produce a verbal description of the ideal variant of the trial concept or sample. Such personal specification can be a highly efficient way of exploring the bases of preference. Indeed, open-ended questioning of the individual has arguably been neglected as a complement to, or even a substitute for, the focus groups used in qualitative market research (Booth, 1988).

Description can only be exploratory, however. Panels' descriptive scores for different sample 'mixtures' have been given far too much weight in sensory evaluation. As we shall see, even the importance that the customer (or the expert group) attaches to certain aspects of the product provides no more than a preliminary indication (Booth, 1987b; Booth & Blair, 1988). No analysis of descriptive scores by themselves, however sophisticated, can quantitate the power with which objective factors influence perceptions of the product, let alone choices made among competing brands.

Segmental positioning

Widespread customisation is unlikely but it has become very evident that there can be good business in targeting brands and sub-brands to distinct segments of the market. A commercially significant minority of consumers may differ from others. They may prefer a different quantity of some factor. They may even have a qualitatively distinct pattern in the influences on their choices.

Longstanding brands of a product often differ in their formulation or marketing attributes, sometimes quite substantially, and yet they still hold profitable fractions of the same market. So, it has seemed sensible to try to design the whole marketed 'mix' of a brand to appeal to a specific region or lifestyle bracket. The brands that this particular offering competes with may appeal more to other segments. Furthermore, such targetting can cultivate an increase in demand and even open up new market positions.

The growth of such product differentiation, market segmentation and strategic positioning has made more obvious the importance of understanding individual customers. It has become harder to sustain the delusion that there is an average or typical consumer. Sales figures can still be used as feedback from undifferentiated consumer ciphers. Nonetheless, the realisation is growing that this alone does not suffice for the most profitable strategy.

Thus, the pressure to find the one optimum formulation is easing. The salient issue now is how variations in the marketing mix relate to the different sorts of 'taste' out there in the market.

Social statistics and individual cognitive science

The assumption that there is a 'perfect' product (i.e. a combination of features that satisfies everybody) was not, however, a freely chosen and rational approach to brand management. On the contrary, it has been a necessity, born of a fundamental limitation in all established methods of sensory evaluation and market research.

All current methods of data gathering and analysis rely in the end on the most unrealistic presupposition that all or most customers react according to one general pattern. This defect persists even in what in recent years has passed for individualisation, both in sensory mapping and in market segmentation research.

Thus, the usual procedures in sensory evaluation are to count the number of differences or preferences evoked from a panel or to average the scores or extract a consensus space. What is called 'individualisation' merely consists of characterising individuals relative to those averages or to the imposed consensus.

It must be added immediately that, in psychology itself, the main research techniques also enforce such unrealistic generalising about people. From fundamental psychophysics to applied social psychology, analyses of grouped data have been used to infer what everybody, or at least most people, might be doing. Paradoxically, this is even true of the tradition in psychology concerned with differences between individuals: the multivariate statistical techniques of classical psychometrics, like those recently developed for sensory evaluation, presuppose that the same variables interact in the same general manner in everybody, and that people differ only quantitatively in their positions within a supposedly universal structure of the human mind.

In some circumstances it may be that everybody is indeed qualitatively identical in pattern of interactions among the influences on a preference. Yet that is an empirical issue. It is ignored by the established methods. Neither univariate nor multivariate analyses of grouped data (variances, ordinal or frequency) open the investigator to the possibility of qualitative differences in causal structure. Behaviour gets treated as mere 'noise' if it is in the opposite direction to the univariate hypothesis or cannot be forced into the multivariate consensus. Quite substantial minorities can be swamped. An average score or a consensus picture may even be artefactual, with most people having characteristics nothing like the numerical middle ground.

This extraordinarily unrealistic approach has been unavoidable because, in the absence of a clear general theory of cognitive integration in the individual, there has been no way to avoid disastrous non-linearities, rendered fatal to analysis by grouping the data. If all or most of the data were monotonic, this problem might not be so great. However, at the heart of the data needed for measuring influences on consumers' choices are functions that are inherently quadratic with idiosyncratic inflections, i.e. individually differing peak preferences. Worse still, preferences (as also sensory differences) can be based on quite different dimensions in different people and some factors may interact in one way for some consumers and in another way for others. As a result of any of these effects, a considerable number of assessors may fit very poorly to the best consensus that can be forced on the data. The region at the ideal-point averages of a constellation of factors may contain very few people. The aggregate response implied by the consensus space could be totally misleading.

The actual causal relationships that concern sensory evaluation are in each individual consumer's mind. They are not in the total vote or the average score that results from many personal decisions of unknown diversity. Investigations must be designed and analysed in ways that allow for these basic realities.

Genuinely individualised analysis of influences on overall choice can be achieved only by ceasing to regard sensory or consumer tests as the collection and statistical analysis of numerical scores according to standardised procedures. The statistics of the voting poll and the multivariate 'fishing expedition', or the P value or average score from a test panel, are insufficient in principle for the causal analysis that is essential to any science of consumer behaviour. Such premature aggregation blurs potentially crucial distinctions between the dispositions of different purchasers.

We need a method that allows for the non-monotonicity of preference and the diversity of interactions among influences on it. Then sensory evaluation could provide scientific evidence about the market impact of the technical features of products. Market research could rectify the similar deficiencies in its guidance to the marketeer concerning the values consumers attribute to the brand. Indeed, research into sensory characteristics and marketing attributes could be integrated, as it must be if we are to obtain reliable information on what customers want. While sensory and consumer data remain separate, technical and marketing decisions about product development can be made only by combining qualitative relationships between the two sorts of information with intuitive impressions of what is really happening in the factory and the market.

The effort currently spent in investigating people's responses to physical and verbal propositions thus needs to be redirected into a thoroughly scientific approach to what is going on in the consumer's mind — that is, measurement of the cognitive integration that controls the individual's brand-choice behaviour. The information yielded by such an approach is bound to be more realistic and less misleading than current analyses. It will also be more readily interpretable for both technical and marketing people and richer and more precise for the same amount of research effort.

THE BASIS OF CONSUMER SCIENCE

The key commercial issue is: what are the causal processes that go on in the individual customer as she makes her choices at the point of purchase? Preference tests, as usually designed and analysed, provide no answer. No amount of preference data by themselves can tell us whether to (and how to) improve a brand or the product formulation going into it. We must relate the preference data to operationalised measures of the attributes of the brand that are actually operative in the mind of each customer making choices.

That is to say, the relevant facts can be ascertained only by using the same basic principles that apply in every experimental science. We need to formulate and test empirical hypotheses about relationships between each customer's relevant assessments and the observable influences on them. Some observables will be physicochemically measurable aspects of the product and its packaging. Others will be verbal and pictorial material cueing the brand's image, whether displayed on the purchased item or attributed to it via pricing, advertising, cultural tradition, etc., in the attitudes that the customer brings to the purchase.

Thus, the properly psychological approach to sensory evaluation should first elucidate each customer's reactions to the inherent characteristics of the product, set within the marketed mix of the brand and the situation in which the customer may choose to purchase or use the item or one of its competitors.

We should not aggregate data until we have calculated the strengths and interactions of the observable and controllable influences on each person sampled. Then this summation of the individuals' objective motivational characteristics across the panel must be done in a way that is open to structural differences between people.

Also, the sampling of the panellists must be tuned to the business aim of the study. This may be only to explore the diversity of consumers, with a view to understanding the sorts of opportunities available. If, rather, the aim is to estimate the response of the market, then of course the panel must be a representative sample. The size of either sort of panel required will be considerably smaller than that needed by the established survey methods, because we are measuring what is actually going on, not making some theoretically unfounded statistical prediction.

Particular attention to sensory factors is justified for those classes of product where purchasing behaviour is likely to be influenced by the inherent features. Usually, such sensory influences will be remembered from past experience of using a brand or its competitors. It is therefore essential that the sensory influences on individual behaviour be measured in adequate simulations of the actual contexts of purchase and use of the marketed brands. Such realistic in-use testing would greatly enhance the role of sensory evaluation in the research guidance to retailers, manufacturers and commodity producers.

Benefits to the technical/marketing interface

The labour of collecting consumer preference data is at present put to use largely to identify likely 'winners' in the market. Nothing more can be done with the data, because they are not interpretable into quantitative specifications for changes in processing, formulation, packaging or advertising, for example. The rationale for a decision to make a change is based on intuitive understanding of the technology and the market, not on data that relate the two in a manner that can be acted on directly.

As things are, therefore, the technical side often bemoan the lack of sensory control in consumer product testing and the lack of realism in the specifications that marketing people generate. Yet equally, for their part, market researchers complain that sensory evaluation, even when related to hedonics, is poorly predictive or even quite misleading as to the performance of the whole marketing mix. Brand values can dominate formulation quality. Certainly the marketed attributes can add to or even drastically modify the preferred sensory characteristics.

A scientific approach would identify all the important influences on customers' preferences and measure the idiosyncrasies of their integration into choices. If each influence were identified by its effect on choice, the data would interpret customer behaviour in terms of the company's production and marketing operations. That is, this customer psychology would provide the technical people and the marketeers with directly usable information on, for example, how to maintain or revive a well-selling brand or how to appeal to those who did not buy it, without detriment to those who already were doing so.

Thus, objective, individualised analysis of influences on consumer behaviour provides a powerful tool for more effective management of the interface between the disparate functions of Marketing and Technical Development. Such research yields a common language and a mutual education for those with different intellectual backgrounds and career objectives. This operationalised consumer intelligence enables separate and even rival organisational structures to be better harnessed to the overall objectives of the business.

Such sensory consumer tests of branded and priced goods at the point of purchase (or, where relevant, during use) are essential also within the laboratory environment, to increase the effectiveness of technical input to product development. Analysis of the contributions of both sensory and image factors to individuals' purchasing preferences is equally vital within the marketing function, by putting brand management onto a sound empirical basis.

INSTRUMENTAL AND PROCESS FACTORS IN PREFERENCE

The principles of this individualised approach to sensory factors in consumer preferences are now outlined. Then some practicalities are considered of extending the current practice of sensory evaluation to accommodate these improvements in the scientific information on which business decisions can be made.

The basic requirement is choice data, person by person, related to their physicochemical and attributional determinants in the context where consumers are making the decisions that need to be understood.

The sensory performance of preference scores

The objectivity of a preference-based approach to sensory evaluation, just like any other approach, lies in its success at relating human responses to observed influences on them, such as may be estimated by instruments. It is not the dictionary meaning of a panel's vocabulary that makes it 'sensory', but whether the descriptors reliably reflect the action of measured or at least controlled factors inherent in the product. Thus, 'thickness' *may* be a sensory term but that is only known if we can closely relate the panel's scores using that term to a measure such as viscosity or perhaps some more complex function of physical parameters (Kokini, 1985). In the same way, then, 'thickness preference' (how much I like it for its thickness) and even overall (i.e. uncharacterised) preference scores are entirely capable of being strictly sensory data, if they can be tightly related to viscosity, etc.

Hence, the identification of objective determinants of preference that are inherent in the product itself is a demonstration that the preference assessment procedure yielding that information is a truly sensory test.

Precision of sensory preferences: linear psychophysics

From the time of Bartlett 50 years ago, the British 'Cambridge School' of experimental psychology has consistently pursued an objective approach to mental processes, albeit relying almost exclusively on grouped data. Cognitive mechanisms are identified from people's performance in tasks designed so that the relationship of the behaviour produced to the information presented distinguishes between theoretical alternatives.

This scientific strategy has yielded much information about the often unconscious processes involved in which is usually termed 'scaling', i.e. extracting quantitative data from human judgments (Poulton, 1968, 1988; Laming, 1986). The practical application of these results (or indeed their use for fundamental theory building) has, however, been very sparse.

This is in large part because gross misconceptions about human data are widespread (Booth, 1987*a*, *c*). These include the interpretation of verbal scores as measurements of subjective magnitudes and the textbook confusion of response formats (such as labelled numbers, lines and boxes) with psychological scales. Such assumptions are gratuitous because, although argued over at great length, they are not submitted to theoretically adequate empirical tests. Sensations and affect are cognitive processes that are much more difficult to identify, let alone to measure, than usually appreciated (Booth, 1987*c*). If, instead, in the first instance, the observed response is treated as a measure of the objective stimulus, questions of how stimuli and responses scale onto sensations can be left until we have the experimental design that can distinguish between the possibilities (Booth, 1979, 1987*a*; Anderson, 1981; Booth *et al.*, 1987).

Furthermore, there are major sources of imprecision in scoring tasks as they are commonly designed. The tasks are often far too difficult and also riddled with distorting influences. This compromises the assessor's ability to scale, i.e. to perform quantitatively. The Cambridge psychophysicist Poulton (1968, 1988), for example, has long pointed to the evidence that quantitative judgments, e.g. sensory scores, are subject to biases from the context in which they are made. The judge's performance is liable to be distorted by wide and uneven ranges of stimuli that are uncoordinated to the range of responses, and by responses themselves that are hard to use linearly (such as two or more decades of numbers, ratios or multiple verbal anchors).

Hence, we may expect the perception and scoring of differences among a set of objects to be accurate only if we avoid such biases. The undistorted performance may then prove to be estimated by a fairly simple mathematical function, as many natural phenomena have turned out to be when adequately understood and measured. That is, properly tested psychophysical functions could well be linear, so long as the physical measure represents the effective stimulus.

Booth et al. (1983) specified a simple set of procedures for minimising psychophysical biases. Because their interest was in food salt preferences, those procedures were especially suited to the measurement of influences on product acceptability, or indeed of any type of motivation, but they can be extended to purely perceptual tasks.

Booth et al. (1983) coupled their methodological proposals with a theory of the effective stimulus. This theory was an empirically testable extension of a speculation attributed originally to Fechner, that intensity judgments are based on discriminations of differences in intensity. The theory they proposed was that preferences (and indeed descriptive judgments) would be based on perceptual differences. On this theory, unbiased differences in preference scores should be proportional (possibly equal) to discriminable differences among stimuli that affect preference. The same theory applies to intensity judgments and to characterised differences from the most preferred intensity. That is to say, all responses affected by a stimulus lie on that one sensory scale, although of course different responses may vary in sensitivity.

This empirical hypothesis, together with the bias-avoiding procedures, provides a measurement model that can be used in the analysis of any graded performance data. If the theory is true and the data have been satisfactorily collected, the quantitation will be very precise, i.e. tight linear relationships will be obtained between levels of an influential stimulus and preference or intensity responses.

Indeed, the model can be turned on its head. The genuinely sensory nature of a descriptor can be tested or the identity of the physical aspect of the product that is evoking it can be ascertained by the precision of the relationship between the putative causal input and the output in question (Booth, 1987*a*). Furthermore, the exact nature of a combination of influences that is determining preference (e.g. overall sensory strength or some complex percept) can be identified by the preciseness of the regression to the integrative response from discriminable differences in the combination of influences on it.

This model should not seem strange or daunting. It does no more than treat the mind as an explicable system. In psychology as in physics, engineering, physiology and economics, the causes of an effect on the output of a system are those aspects of the input that have most impact on the output.

THE CAUSAL STRUCTURE OF PREFERENCE

Thus, for example, the sensory determinants of brand preference are those physicochemical aspects of the brand that make most difference to whether it is chosen or not.

It is crucial to analysis of the causal structure of such choices to appreciate that all normal adult motivation to interact with familiar objects is acquired by experience of their complexity. For the case of foodstuffs, it has been demonstrated experimentally that familiarity or associative experience establishes a particular stimulus configuration as the most motivating situation (Booth *et al.*, 1972; Booth, 1981, 1985). This mechanism for appetite or acceptability was presupposed when formulating the linear model of preference psychophysics originally (Booth *et al.*, 1983).

It follows from the learning of acceptability that any influence on choice of a brand has a peak level for the individual consumer in the multidimensional context of the competing brands and the choosing situation.

Peaked preferences

The learning processes (such as behavioural habituation, stimulus conditioning or acquisition of discriminative stimulus control) will ensure that preferences for characteristics that are usual in a product will have a peak value for each characteristic — its ideal point for that person in that situation. This is because overall liking will be reduced if any stimulus element contributing to the ideal configuration is increased or reduced in amount from the learned level.

On the same theory, however, a foreign characteristic will usually reduce preference monotonically. There will not be a triangle of tolerance but simply a decrease in liking the more there is of the 'taint'. One might consider that there is still an ideal point, but it is at the zero level of the taint stimulus dimension.

In some other cases, the preference peak can be at or beyond the highest conceivable level of the influence on acceptability. This occurs with some attributes that are conceptual and unlimited, such as healthfulness. It would also occur with an appetising effect that is innate. Perhaps the only example is sweetness. Because learning incorporates sweetness preference in the ideal configuration, the innate reaction can only be expressed in an unfamiliar context (Booth et al., 1987). Unflavoured sugar solution is not a drink or food for most of us and so may evoke a monotonic increase in pleasantness with greater sweetness (Booth & Shepherd, 1988). Children may be introduced to novel foods by this means and then acquire a liking for the extra sweetness (Booth, 1987d).

Another important aspect of learned motivation is the likelihood that, whatever characteristics are influential, their most preferred levels are likely to be personal and the interaction of determinants of choice may be idiosyncratic. For, no-one's history of experience with a product type need be the same as anybody else's. Hence, the structure of each person's tolerance of deviations from an ideal configuration can be ascertained only by causal analysis of that individual's observed performance in the relevant situations. A set of people cannot be represented realistically by generating a peak preference from the grouped data.

The tolerance triangle

It also follows from the learning of preferences that any particular influence on choice will form an isosceles triangle. Learned responses decrease in strength the more different from the learned stimulus that the test stimulus seems (an effect known to behaviourists as the 'stimulus generalisation decrement'). That is, equal decreases in acceptability will occur the more discriminable the presented level of the influence is on either side of its ideal level for that person in that context. The influence is identified by the linearity of its relation to choice and, in the case of a peaked function, linearity means the same slope on each side of the apex.

Again, the model can be reversed in order to analyse data. The graph on the high side of peak preference can be reflected in the ideal line, i.e. the function can be unfolded (Coombs, 1964). Linear preference psychophysics will give a straight line from insufficient to excessive levels of the influence. The measurement model of Booth et al. (1983) predicts such straight-line unfolded preference functions, so long as unbiased preference responses are plotted against the influence on preference in units of equal discriminability.

By implication, the asymmetrical and rounded inverted-U shape usually observed in hedonic functions is artefactual. Asymmetries often arise from a biased set of stimuli and/or incorrect scaling of the objective influence on preference. Rounding of the peak can arise from the grouping of data from a set of individuals whose ideal points are at best normally distributed. Also, however, even an individual's observed preference peak can be rounded. This must be expected when there is something 'wrong' with all the samples other than the factor whose role in preference is being plotted. The peak must then be truncated because there is an upper limit on overall preference even at and around the ideal level of the influence on preference being tested. Nevertheless, the lower levels of acceptability that were observed can still be interpolated through the discontinuity around ideal (created by unfolding a truncated peak). This co-linear interpolation provides an estimate of the unobserved sharp point of maximum preference for the factor under test (Booth & Shepherd, 1988). In this way, it can be feasible to optimise some factors while others remain suboptimal throughout testing.

STRENGTH OF AN OBJECTIVE INFLUENCE ON PREFERENCE

This psychological mechanism for any discrete influence on choice, when generalised for all effective influences, provides a multi-dimensional psychophysics of consumer acceptability. Analysis of the unfolded tolerance triangles in each consumer's data builds the missing bridge between sensory and hedonic data and between knowledge of the technical and marketing factors that the company puts into the brand and of the aggregate and individual responses of consumers to that and competing brands.

The tolerance line

The equation of an unfolded tolerance triangle can be estimated from the data by linear regression. This psychophysical function has the three logically independent characteristics of any regression line. These are the slope, an intercept, and the residual variance after extracting the linear trend.

Each of these characteristics of psychophysical functions (although usually in group data) has been used as a measure of sensitivity to the stimulus. The importance of sensitivity is that it is also the strength of the effect of the stimulus on the response that has been recorded. Thus, the sensitivity of choice ratings or behaviour to a factor in the brand is the power of that influence on choices in the customer and context tested.

Intercepts

Most attention traditionally has been given to the intercept with zero intensity, i.e. the absolute threshold level of the stimulus. Even in some practical contexts, this is still sometimes taken to be the best measure of sensitivity. Yet a measure at such low levels of the stimulus is entirely irrelevant for stimuli whose practical effects are at suprathreshold levels. Some have assumed that threshold and suprathreshold sensitivities will be related but there is no foundation for this notion in theory or in observations. Indeed, Laming (1986) has shown that absolute and differential detection tasks are performed in quite different ways.

It has seldom been appreciated that there is a psychological zero point in any peaked preference function. This zero is the ideal point. There, the individual has absolutely no problem about the level of that influence, e.g. coffee-bean roasting time or sugar concentration. That is, on this (unfolded) scale, insufficient levels of the influence are below the zero (negative) and excessive levels are above zero (positive). In terms of the measurement model, a sensory factor may be so many discriminability units below or above ideal.

This then is an intercept measure of the strength of a preference factor — how much of it is ideal. If you need just a soupçon, the factor might be regarded as a strong influence on your preference, whereas something that has to be ladled in is a weak motivator. Be that as it may, the ideal point is clearly a very important parameter of preference, and it is relevant for monotonic preference functions as well as for peaked ones.

There are two other intercepts which are no less important psychologically in the scaling task. They can be of considerable practical significance too. These intercepts arise from the other anchor point in any scoring of choices, e.g. the point of rejection ('never buy'). Unfolding a peaked preference function yields two rejection points ('too much' and 'too little'), although a one-sided preference (such as taint) has only one.

There are several problems with a rejection-point(s) parameter, however. It relates to situations that may be less familiar than near-ideal situations, and so memory may provide a less accurate stimulus value for comparison. Furthermore, one would hope that the attainment of maximum appeal is more relevant to product development than the avoidance of total lack of appeal. Finally, there are two rejection points for a peaked preference function but the unfolded straight line has only one intercept parameter to be assigned. A convenient solution to this degree of freedom problem is to use the distance between the rejection points as a measure of slope and keep the ideal point as the intercept. The values of both rejection points can be deduced from this slope and the ideal point. The ideal will be midway between the rejection points, since the data fit the isosceles triangle model of tolerance.

Slope

Slope (i.e. the exponent if there were a power function) is now the common measure of suprathreshold sensitivity in psychophysics. Similarly, proportion of the dynamic range is a common measure of sensitivity in engineering (Truxal, 1972).

An unavoidable and severe limitation of a slope parameter is that it totally confounds sensitivity with response style. The numerical value of a slope depends on how the assessor interprets and uses the response categories, be they 'magnitude estimates' or ratings anchored on two or more categories (whether of intensity or of belief, liking, intention or disposition to choose). The slope of a psychophysical function extending from threshold to saturation may be less confounded by response style but the asymptotes are critical and difficult to estimate accurately and in any case such data ranges are not of practical significance to product development.

A consideration of slopes brings out the general requirements for good scoring performance. There is insufficient appreciation of the practical implications of the fact that two points define a line in any space. If an investigator is rash enough to provide more than two anchors (whether words or numbers, or the whole integer series implied by the single numerical anchor used in ratio rating), there may be a different slope between each pair of categories. Two and only two anchor categories should be presented in any quantitative judgment task.

The two anchors should also be highly usable. Otherwise, they will not be very effective at specifying the dimension on which the assessor judges. Thus, for example, one person's extremely sweet, one-tenth as sweet, or as sweet as that is loud or long, bears no determinate relation to another person's and is therefore all too vague for the assessor using such anchor wordings. For preference scoring, like and dislike or 'too much' and 'too little' have no definite logical connection with any particular behaviour. Action categories are more determinate behaviourally, e.g. 'always choose' and 'never choose'. Even so, two people in the same situation do not have to have the same relationship between their judgments that they would never choose and their rejection behaviour. Disposition to accept or reject can be verbally exaggerated, as can the criteria for weakness or strength of a percept.

Consequently, slopes are of only limited value as a measure of the sensory sensitivity of a consumer's preferences, which is how they were used in our first reports on preference psychophysics (Booth et al., 1983; Griffiths et al., 1984). We limit our use of them now to where rejection-point values are needed as well as ideal point values (Conner et al., 1988a, b). They can, nevertheless, prove to be as good or better than a stricter measure of sensitivity at explaining overall choice (Booth & Blair, 1988).

Residual variance

Residual variance is a more satisfactory measure of sensitivity than slope. It has been neglected in psychophysics, although at least one advocate has now appeared (Weiffenbach, 1989). Yet the correlation or regression coefficient (residual variance as a fraction of total variance) is widely used elsewhere as a measure of the strength of a statistically predictive relationship.

Where the direction of causation is known, as in a psychophysical experiment, the coefficient is a measure of the strength of the evidence for causality in those data. However, the regression coefficient or its attached P value is not a measure of causal strength as such; it is merely a statistical fact about the data (MacRae, 1988). If we wish to measure how sensitive the response is to the stimulus, or what psychological distance there is between different stimulus levels, then the data must be used to estimate that characteristic of the psychological mechanism involved. To such a measure we now turn.

Residuals relative to slope: performance sensitivity

A mechanistic measure of the sensitivity of a set of psychophysical judgments can be obtained from the relationship between the residual variance in the responses and the slope of the psychophysical function.

In the simplest case (which, thus far, bias-minimised individual data have always proved to be), elementary signal-detection theory (SDT) can be applied. This is Thurstone's scaling Case V, or its mathematical equivalent — the classical calculation of the just-noticeable-difference

(JND) or difference 'threshold' at any particular stimulus level. This case applies to a graded performance that shows constant residual response variance over the range of the observed data lying on a linear psychophysical function. When performance meets those conditions, then a simple formula dividing response variance by slope can estimate, for anywhere in the tested stimulus range, the difference in stimulus levels that corresponds to 50%-correct discrimination, i.e. one JND. The equation is:

$$\text{JND} = \text{antilog}\,(2 \times z\{25\%\} \times \text{SD}/m)$$

where $z\{25\%\}$ is the z score for 25% of each response distribution for two stimuli one JND apart, SD (standard deviation) is the square root of the mean residual response variance and m is the slope of the linear regression (Conner *et al.*, 1988a, b). This formula gives the JND as a ratio to the stimulus level above which it is discriminated. This 'threshold' stimulus-step ratio (e.g. 120%),

$$\text{JND} = 1 + \text{WR}$$

where WR is the Weber ratio (e.g. a 20% increase is just discriminable). In fact, this calculation assumes no threshold of discrimination. It simply sets as the unit of discriminability a value of about unity (0·96) for the SDT response-bias-free sensitivity parameter (d').

This calculation is not merely a mathematical model forced on the data (Thurstone, 1927), let alone is it a statistical fix. It does not assume that unbiased ratings should be proportional to a cumulative JND scale (McBride, 1983), that the Weber ratio is constant or that psychophysical functions must follow a semilog law. The calculation derives from a fundamental mechanistic theory and the theory can be tested on the data to which the model is being applied. This fully empirical hypothesis is incompatible with some of the mathematical models of choice but is highly consonant with recent very general models of recognition and multidimensional differences (Ashby & Perrin, 1988; Ennis *et al.*, 1988).

The causal mechanism here hypothesised to underly all psychological performance is the objective behavioural version of Fechner's introspectionist speculation that the discrimination between two sensations and the strength of a sensation are based on the same process. On this hypothesis, the influence on intensity rating or choice is identified as being the phenomenon that gives a constant Weber ratio for the response being observed. This is a scientific solution to the problem of deciding which are the correct physical units for a psychophysical function: if stimulus ratios do not give a straight line against unbiased responses, then we are not using the stimulus measure that the assessor's judgments are using. Of course, a stimulus ratio scale can be expressed in the equal-difference units convenient for graphs by taking logarithms. Thus, this mechanism predicts the Fechner-Weber semilog 'law' of psychophysics but only for appropriately designed and analysed tests of an individual's performance.

When we put together such measures of causal strength for all the influences on an assessor's response, we are in effect converting a multidimensional psychophysics into quantitative cognitive science. This is done very precisely for the mind of an individual, because that escapes the noise and fallacies of grouping data that afflicted earlier efforts to establish a 'cognitive algebra' (Anderson, 1981), currently being extended to psychophysical issues (McBride, 1986; de Graaf *et al.*, 1987). Any fuzziness in the individual analysis is a precise description of the noise in the observed performance with respect to the influences as measured.

This JND parameter for sensitivity of an individual's performance, unlike slope or residual variance alone, is not affected by the quantitative meaning given to the response categories. It reflects only how the person manages to use the meaning of the words to express perceived differences between samples. Hence it is a measurement of psychological distance that is independent of response style.

This measure of the strength of influence of a stimulus on an individual is applicable to any response, including preference scores (unfolded if peaked). However, since the sensitivity of a preference to deviations from ideal may be less than what is discriminable, i.e. it is what is tolerated, the JND from a preference function is better called a 'just tolerable difference' (JTD). We have dubbed the corresponding Weber-type ratio a 'tolerance discrimination ratio' (TDR).

The TDR can be used in conjunction with the ideal point to estimate a range over which an assessor would not react adversely to variation in the brand. Only beyond this ideal range, on either side, is there any real chance that her or his preference might begin to decline.

This then is a system for translating levels of sensory factors in product designs into the dispositions of individual consumers to accept or reject a variant of the whole marketing mix in the tested situation.

Aggregation

The practical interest is in how diverse these dispositions are, not only in ideal points but also in how differences from ideal in different dimensions interact.

The individual mechanistic analysis identifies and measures qualitative differences in the patterns of determinants of consumers' behaviour. The prevalence of each type of such 'psychographics' in any segment of the market can be estimated by representative sampling in adequate numbers. The aggregated effect of whatever diverse or homogeneous dispositions exist can be estimated simply by summing the individuals' data from a sample representative of the market of interest.

MEASUREMENT OF OBJECTIVE SENSORY PREFERENCES

We could coordinate the results of preference tests to business planning much more effectively if we used methods that did not rely on the emergence of one clear 'winner' or an overwhelmingly popular pattern of influences on choice. Many of those voting for the leading brand may find it no more than the best compromise, with the alternatives still worse. It would be much more useful if we could find out how winners of panel votes (and some decent losers) fall short of perfection, person by person in our panels. Clearly, only something like the above approach can tackle this issue realistically and cost-effectively.

The strengths of influences of objective sensory factors in consumer preferences are best estimated by the use of the above cognitive theory of preference in what might be called the Controlled Preference Test.

Tested state and situation

Adequate control requires assessors to be in the state in which they would show the actual behaviour to be predicted. For example, if trial purchase choice is to be estimated, then one panel of assessors should represent those who will have seen the advertising for the new product and so be shown something like it before testing, and both this pre-aware and an unalerted panel should be presented with the sort of shelf display likely to be used by the retailer. If effects of use on repurchase are of interest, then assessors should be users of that type of product and the test situation should adequately simulate or recall one of their major uses of the type of product to be evaluated.

Consumer panel numbers

Where the fundamental scientific method of varying putative causes and observing the effects of interest is logistically practicable, then the richness and precision of the data will considerably reduce the number of consumers who have to be tested.

This is because the size of the panel needed is not determined by the social statistics of the opinion poll. The important consideration in the first instance is that the panel samples the diversity among potential customers. Only if there prove to be substantially different psychographic types or if the precise distribution of ideal points is important do further questions arise. It then matters whether the segment actually sampled is representative. It may then be necessary to estimate the prevalences of the different preference patterns in different segments to see whether one of them is common enough for a financially viable positioning strategy. If the commercial tolerances on such a prediction are tight, then a considerably larger and thoroughly representative consumer panel will need to be measured individually on the key factors. Nevertheless, this fully quantitated panel is unlikely to have to be anywhere nearly as large as the pre-test market panels used with existing aggregate models to predict market shares.

Even more important, the panel aggregate involves no consensual statistical calculations. It is merely the sum of individualised analyses. So it is simple arithmetic to break back subaggregates to any segmentation criterion wished.

Product sample selection

The greatest precision of diagnosis of the structure of an individual's preference is achieved by controlling the characteristics of the presented samples so that each influence has a linear effect, and also so that different influences are not confounded (and thus poorly distinguishable in their impact).

Often, however, nothing is known about an individual's preference before testing starts. Furthermore, the functioning of the brand may be so poorly understood in the absence of individualised analysis that little can be safely assumed either about most individual consumers. The scientific approach to such ignorance is to frame hypotheses and collect data that test them. The most economical first step is to see how far the most popular version of the product is from the consumer's personal ideal. Whether it is close or distant, subsequent samples need to be varied so that the individual's preference score changes substantially, up and down.

The second sample should test the hypothesis that one of the common influences on preference is indeed important for that consumer (Booth, 1987b, c). If the preference for the first sample was low, the best hypothesis to test is that a likely influence was too low or too high in level. One direction may be more plausible than the other, from what we know of the brands available or from the impression the investigator has of the common preference patterns (even by introspection!). Intuition can also suggest how large a change in level to try in the second sample. If, on the other hand, the first sample did indeed prove to be close to ideal, then the change in the second sample should be substantial and could be either up or down in level of the influence to be tested.

Modest changes should also be made in other likely major influences on preference, to minimise confounding of influences in the data. The inclusion of actually marketed brands in the series of tests might achieve this. The variations among marketed samples might, however, confound some physically independent factors. If these factors may vary substantially in their impact on some people, one or two samples should be prepared that disconfound those variations.

By good fortune the first two samples might indicate the possibility that the two levels of the first-tested factor lie on either side of the assessor's ideal point. Failing this, the third product sample, or at least one before the end of testing, should seek such a result, i.e. should test the hypothesis that the change could be in the excessive direction as well as in the deficient direction. To minimise range bias, these excursions from ideal should be to approximately equal extents.

The third or fourth sample should test another likely factor in preference by changing its level markedly, and so on through other less likely influences for as long as testing can continue in a session or over two or more sessions.

If the first large change appears not to affect preference, the risk can be taken of moving to another factor in the next sample. If a second substantial influence is indicated, then interaction with the first factor should be tested as soon as feasible (possibly in the next product sample), for the sake of disconfounding the two influences; that is, the levels of both factors should be markedly changed. It is probably best initially to assume an additive type of interaction, with for example each sample that contains high (or low) levels of both factors being paralleled by a sample containing a high and a low value.

Four samples are the minimum needed to interact, disconfound and bias-minimise two influences on preference. Eight samples, correctly designed for the individual (which may not be achieved at first try), would cover three influences. Six would be sufficient if no direct test for three-way interaction was made and some modest confounding could be tolerated.

The above principles of sample selection require some measurements of the factors likely to be influencing the customers under test. These measures can be made with laboratory instruments but do not need to be. Indeed, if an influence is poorly understood, a descriptive score might be more useful. Something like labelled price or kilojoules per food portion is its own measure. Expert scoring on consensus descriptions can be used or, better still, the consumers' own scores on vocabulary that they have chosen independently, to the extent that the wording agrees or scores correlate highly across brands. Such measures may confirm that brands already on the market would serve well as the sample set from which each consumer's test samples are selected. Nevertheless, intermediate samples can be helpful in testing very intolerant consumers. As already stated, special samples would be needed if confounds are to be challenged.

Analysis

The data from each individual assessor should then be analysed separately by multiple regression onto preference scores of the distances of factor levels from postulated ideal points. The ideal point for a factor can initially be taken to be the tested level that give the highest (mean) preference score. However, a more refined estimate can be obtained by optimising the ideal point value for the tightest regression of that factor onto preference.

Multiple regression using a partial for each measured factor is the simple extension of the tolerance triangle theory to two or more discrete causal influences. However, factor interactions should also be partialled into the regression. Preferably these interaction terms should be based on some psychological hypothesis. There are often theoretical or empirical reasons to expect a particular form of cognitive integration, such as sweetness or creaminess compensating for bitterness in coffee drinks. To maintain the diagnostic power of this hierarchical multiple regression, an interaction term must not correlate substantially with any of its component simple terms.

The ultimate test of our understanding of an individual's preference structure is consistency in that person's cognitive algebra within a situation. This would be shown by the replication of a tight prediction of choices by a formula that combines all the major influences on preference. A purely sensory parallel to this would be a successful prediction of the rated intensities of a complex texture by a formula combining several independent physical factors, such as one of those proposed by Kokini (1985). An example of integration of a sensory factor and an image factor into preference is an additive interaction between sweetness and labelled caloric content of the sweetener in a beverage (Booth & Blair, 1988).

Bias-permitting tests

If the test samples are tuned to individual preference in the above fashion, the results are highly reliable and exactly interpretable. Nevertheless, it is practicable to run the individualised analysis on data collected by the usual preference-test procedure of presenting a conveniently available set of existing brands and new propositions, regardless of individual differences in tolerance ranges and preference structure.

The individualised analysis of such data must be used to monitor the defects they are likely to have. One of the most serious defects is a bias arising from an imbalance of samples on one or other side of ideal for some assessors (Booth *et al.*, 1986). Indeed, some preliminary individual analysis could be used to decide whether to add or delete samples that frequently contribute to major biases in individuals. These biases will not only reduce precision. More seriously, they will shift the estimated ideal points to extents that are at present hard to quantitate. Such effects have grossly misled optimisation procedures in the past (Booth, 1988). Such distortion might be reduced in aggregates from individual data by stratification for biases in opposite directions, but even that will not be possible if the set of samples biases virtually everybody in the same direction.

PREFERENCES IN SENSORY TESTING

It has always been considered very difficult to relate preference data to instrumental measurements or even to sensory descriptions. Indeed, these difficulties fostered a strong tradition of keeping sensory testing quite separate from hedonic or affective testing.

Nowadays, however, it is widely appreciated that sensory evaluation cannot serve business effectively without using panels of consumers. As a result, sometimes both sensory and hedonic questions are put to the same consumer panel and the panel's answers (or the answers from separate sensory and consumer panels) are interrelated by multivariate statistical analyses.

Selection but not training

Nevertheless, an essential role is still seen for the selected trained sensory panel. It is thought that untrained consumers cannot be effective in complementing instrumental measurements. Trained panels are considered to be necessary to back up quality control, for sensory matching after a change in formulation has been required, to provide systematic descriptive analyses of technical factors, and even for 'preliminary' optimisation of new products.

The contention here is that, on the other hand, directly measuring the determinants of choice in consumers is the only reliable way to investigate taints, mismatches, suboptima and both the describable and the subconscious mental processes by which company-controllable factors influence the market.

Valid business guidance requires the set of people who are tested to be representative of the distribution of ordinary customers in the market under consideration. Company employees are unlikely to be typical consumers. Even consumers from an outside panel cannot represent more than the locality and lifestyle from which they are recruited. Training in sensory discriminations, vocabulary and standards is likely to adapt panelists' preferences. Extensive tests of one type of food product by untrained consumers are liable to sophisticate their palate unrepresentatively.

For purposes of identification of an unknown sensory factor in preference, however, it is more efficient to select untrained consumers whose preferences and hence their perceptions are sensitive to that factor. If the prevalence of such sensitivity is kept a separate issue, then such selection does not invalidate the sensory conclusions.

Representative consumer panel

The first requirement for a fully scientific psychological approach to existing sensory evaluation techniques is therefore that the testing be done on a sample of ordinary consumers who use the type of product under study. Even if quantitative representativeness is not needed, the panel should still be a diverse set of users.

Sometimes the issue is the identity or source of the chemical constituents or physical characteristics that are causing a sensory problem. Even in such situations, what matters to the business is the factor that influences customers' choices. The preference response of an affected consumer can be as sensitive as the descriptive or difference scores of a trained sensory panel (Conner et al., 1986, 1988a; McBride & Booth, 1986). It is therefore foolish to confine sensory testing to differences or descriptions. It risks irrelevance when preference measures of objective differences are available at no greater expense. In fact, descriptive responses are not needed at all, unless preference-relevant instrumental measures are unavailable.

Market-realistic test situations

Sensory preference tests must be carried out in circumstances that have been shown to be an adequate simulation of the context relevant to the business issue(s) being addressed. If a complaint is being investigated, the usage that is likely to have given rise to the problem should be evaluated. If a 'best before' dating is to be decided, it is important to determine not only the time it takes for preferences to decline after common domestic use or abuse of the product, but also the effect on customers who note the days between purchase and the packet date.

Finally, all sensory evaluation should allow for the fact that brand image and purchaser expectations can greatly modify the sensory influences on use, preference or description. At the very least, any conclusion of major importance from unbranded tests should be checked for branded samples in a small panel.

IMPROVEMENT OF EXISTING TECHNIQUES

Established types of sensory testing can be seen as different sorts of weakening of the Controlled Preference Test outlined above. The individualised choice-difference analysis can often be grafted onto an established system, resulting in a striking increase in understanding, precision and development efficiency.

Optimisation

The preference testing already described is the fully scientific approach to adjusting the formulation of a product to appeal to consumers as strongly as possible. The aggregate of the individuals' tolerances or ideal ranges that is best predictive of preference can be interpolated to any proposed formulation to predict the proportion of the population represented who will accept or reject that formulation (Booth & Blair, 1988; Conner et al., 1988b).

Clearly the author must disagree even with fellow-psychologists when they advocate testing of a wide range of stimuli regardless of biases on individuals, using hedonic scoring language and laboratory test situations far from the real-life behaviour that matters commercially, and analysing the aggregated data instead of making individual estimates and summing them.

The purely statistical approach to optimisation becomes extremely complex and is scientifically quite unproductive. Each set of descriptive and preference ratings has to be fitted by high-order polynomial regressions. As well as being plagued by biases and grouping artefacts, the results are purely empirical and add little to understanding of how the product works.

Simplex optimisation is perhaps closer to the approach proposed here. It has the advantage of keeping near to the optimum in the final stages and so reducing the amount of bias that the one-stage, wide-range approaches risk. However, it achieves economy of data only by using differences between individuals to estimate variance. So it is subject to all the fallacies of grouping data that individualised preference analysis avoids.

Profiling relative to ideal

Standard descriptive scoring (profiling) techniques, whether the vocabulary is imposed, standardised across the panel or chosen by each individual, cannot measure the importance of a sensory factor. The discriminative power of a sensory term could be estimated from the ratings and instrumental measures if the data were sufficiently orderly to permit group or, better, individual calculations along the lines of the Weber ratio. This would estimate that objective factor's perceptual importance in the panel test. Yet it would still say nothing about its strength or its role in the structure of sensory influences on consumer behaviour. That is the real issue.

Descriptive sensory scores relative to ideal, on the other hand, can provide some behaviourally relevant information. It is no different to carry out ideal-relative image scoring and what is in fact attitude assessment in the individual. The results of ideal-relative ascription are still only exploratory, however. What people *say* is influencing their choices cannot provide reliable quantitation of how the product actually works in consumers' minds to organise their behaviour. The main use for profiling, for which it should be relative to ideal, is to guide the design of the controlled testing of objective factors in uncharacterised preferences (Booth, 1987b, c, 1988; Booth et al., 1987), as described above.

Indeed, relative-to-ideal sensory profiling with individual vocabulary and analysis is a psychologically quantitative form of qualitative market research, focusing on the factors of interest to the technologists. Such exploration of ideal-relative profiles is least misleading when done in untrained consumers, using each individual's own words.

Difference tests

Difference testing should be regarded as a confession of total ignorance about the psychology of a problem with the brand. The scientific investigation of the unknown is far more difficult than that of the known. Thus, far from being easy and safe to run and to interpret, difference tests are the most difficult of any sensory evaluation technique to use effectively.

As might by now be anticipated, the suggestion here is that such a lack of quantitative and even qualitative information about the problem is best rectified by using a varied sample of individual consumers to provide their own descriptions of the influences affecting preference in the two products to be compared, and then calculating estimates for each assessor of the sizes and directions of the ideal-relative differences described.

Uncharacterised overall choice scores should be obtained for each sample. They may well be as sensitive as any descriptive ratings. They can also be used to crosscheck what each assessor says is the way she or he chooses with the relationships between objective measures and the preferences (Booth, 1988; Booth & Blair, 1988).

The sensory precision of ideal-relative descriptive ratings (Conner *et al.*, 1986, 1988*a, b*) shows that the ideal value of an aspect of a familiar type of product is an internal standard that can be constructed from memory rather accurately. In experiments more closely related to difference testing, furthermore, a sensory panel was as sensitive to differences between test samples and the panellists' individual ideal points as to differences from a physical standard (McBride & Booth, 1986). That study was based on judgments of overall strength of taste of a beverage and so the comparisons of two physical samples involved tasting in succession and relied on short-term memory. Simultaneous visual comparison, for example, might prove to be more sensitive than comparison with ideal. Nevertheless, many visual comparisons (such as between brands on the supermarket shelf) in practice involve glancing from one stimulus to the other and actual sensory juxtaposition is rare in consumer behaviour. Hence reliance on short-term memory is not peculiar to taste testing and the McBride and Booth result probably has general significance.

This emphasises the 'monadic' nature of uncharacterised or characterised choice-difference testing. Hence, the multiple samples usually presented in a difference test can be used to get repeat scores from each assessor. These replicate data can be used to estimate by the TDR calculation the size as well as the direction of the importance for each individual of the difference between the standard samples and the test samples.

Quite often, furthermore, the business issue that difference testing is really meant to resolve is not so much the difference of the test samples from top-quality standard samples as the difference from products as generally bought in the current market. How many current purchasers might think that the test product is tainted (or might actually prefer the extra flavour!)? How long can the product be stored or stay open on the kitchen shelf before consumers think it is 'off' under current market conditions?

On the theory of learned motivation (Booth, 1985, 1987a), individuals' current ideal points are based on their lifelong experience of the market. Hence, shelf-life criteria, for example, could be set up on the proportion of consumers who rated the stored sample as one JTD or more from ideal or from what they expect from the brand. No fresh sample need be presented alongside the stored sample, especially if a comparable new subpanel was used at each length of storage of one batch. Difference testing is of the essence for shelf-life research but it is the acceptability difference (JTD) that needs measuring, not the JND (McBride, 1980), and the difference in acceptability from what customers want, not from an unaged sample.

Thus, accept/reject ratings of at least two samples of the suspect product can be used to estimate in TDR units its distance from the marketed standard as recently experienced by that individual consumer. The spread of these scores across a panel of likely purchasers and users could be used, against a criterion of prevalance of, say, one or two TDRs from ideal, to decide sell-by and best-before dates. The scores could justify changes in production or distribution to reduce a known source of taint. They could also show the need for further research to identify the aging or tainting process. The most effective research technique for this would of course be psychological measurement of the objective determinants of individuals' real-life choices.

REFERENCES

Anderson, N. H. (1981). *Foundations of Information Integration Theory*. Academic Press, New York.
Ashby, F. G. & Perrin, N. (1988). Towards a unified theory of similarity and recognition. *Psychological Review*, **95**, 124–50.
Booth, D. A. (1979). Preference as a motive. In *Preference Behaviour and Chemoreception*, ed. J. H. A. Kroeze. Information Retrieval Ltd, London, pp. 317–34.
Booth, D. A. (1981). Momentary acceptance of particular foods and processes that change it. In *Criteria of Food Acceptance: How Man Chooses What He Eats*, ed. J. Solms & R. L. Hall. Forster, Zurich, pp. 49–68.
Booth, D. A. (1985). Food-conditioned eating preferences and aversions with interoceptive elements: learned appetites and satieties. *Annals of the New York Academy of Sciences*, **443**, 22–37
Booth, D. A. (1987*a*). Cognitive experimental psychology of appetite. In *Eating Habits*, ed. R. A. Boakes, M. J. Burton & D. A. Popplewell. John Wiley, Chichester, pp. 175–209.
Booth, D. (1987*b*). Individualised objective measurement of sensory and image factors in product acceptance. *Chemistry and Industry (London)*, (1987, Issue 13), 441–6.
Booth, D. A. (1987*c*). Objective measurement of determinants of food acceptance: sensory, physiological and psychosocial. In *Food Acceptance and Nutrition*, ed. J. Solms, D. A. Booth, R. M. Pangborn & O. Raunhardt, Academic Press, London, pp. 1–27.
Booth, D. A. (1987*d*). Evaluation of the usefulness of low-calorie sweeteners in weight control. In *Developments in Sweeteners — 3*, ed. T. H. Grenby. Elsevier Applied Science, London, pp. 287–316.
Booth, D. A. (1988). Practical measurement of the strengths of actual influences on what consumers do: scientific brand design. *Journal of the Market Research Society (UK)*, **30**, 127–46.
Booth, D. A. & Blair, A. J. (1988). Objective factors in the appeal of a brand during use by the individual consumer. In *Food Acceptability*, ed. D. M. H. Thomson. Elsevier Applied Science, London, pp. 329–46.

Booth, D. A. & Shepherd, R. (1988). Sensory influences on food acceptance — the neglected approach to nutrition promotion. *BNF Nutrition Bulletin*, **13**(1), 39–54.

Booth, D. A., Lovett, D. & McSherry, G. M. (1972). Postingestive modulation of the sweetness preference gradient in the rat. *Journal of Comparative and Physiological Psychology*, **78**, 485–512.

Booth, D. A., Thompson, A. L. & Shahedian, B. (1983). A robust, brief measure of an individual's most preferred level of salt in an ordinary foodstuff. *Appetite*, **4**, 301–12.

Booth, D. A., Conner, M. T. & Marie, S. (1987). Sweetness and food selection: measurement of sweeteners' effects on acceptance. In *Sweetness*, ed. J. Dobbing. Springer-Verlag, London, pp. 143–60.

Coombs, C. H. (1964). *A Theory of Data*. Mathesis Press, Chicago.

Conner, M. T., Haddon, A. V. & Booth, D. A. (1986). Very rapid, precise measurement of effects of constituent variation on product acceptability: consumer sweetness preferences in a lime drink. *Lebensmittel-Wissenschaft und -Technologie*, **19**, 486–90.

Conner, M. T., Booth, D. A., Clifton, V. J. & Griffiths, R. P. (1988*a*). Individualized optimization of the salt content of white bread for acceptability. *Journal of Food Science*, **53**, 549–54.

Conner, M. T., Haddon, A. V., Pickering, E. S. & Booth, D. A. (1988*b*). Sweet tooth demonstrated: individual differences in preference for both sweet foods and foods highly sweetened. *Journal of Applied Psychology*, **73**, 275–80.

de Graaf, C., Frijters, J. E. R. & van Trijp, H. C. M. (1987). Taste interaction between glucose and fructose assessed by functional measurement. *Perception & Psychophysics*, **41**, 383–92.

Ennis, D. M., Palen, J. J. & Mullen, K. (1988). A multidimensional stochastic theory of similarity. *Journal of Mathematical Psychology*, **32**, 449–65.

Griffiths, R. P., Clifton, V. J. & Booth, D. A. (1984). Measurement of an individual's optimally preferred level of a food flavour. In *Progress in Flavour Research*, ed. J. Adda. Elsevier, Amsterdam, pp. 81–90.

Kokini, J. L. (1985). Fluid and semi-solid food texture and texture–taste interactions. *Food Technology*, **39**(11), 86–92, 94.

Laming, D. R. J. (1986). *Sensory Analysis*. Academic Press, New York.

MacRae, A. W. (1988). Measurement scales and statistics: What can significance tests tell us about the world? *British Journal of Psychology*, **79**, 161–71.

McBride, R. L. (1980). Can shelf life be measured? *CSIRO Food Research Quarterly*, **40**, 149–52.

McBride, R. L. (1983). A JND-scale/category-scale convergence in taste. *Perception & Psychophysics*, **34**, 77–83.

McBride, R. L. (1986). Sweetness of binary mixtures of sucrose, fructose and glucose. *Journal of Experimental Psychology: Human Perception & Performance*, **12**, 584–91.

McBride, R. L. & Booth, D. A. (1986). Using classical psychophysics to determine ideal flavour intensity. *Journal of Food Technology*, **21**, 775–80.

Poulton, E. C. (1968). The new psychophysics: six models for magnitude estimation. *Psychological Bulletin*, **86**, 777–803.

Poulton, E. C. (1988). *Bias in Quantifying Judgements*. Lawrence Erlbaum Associates, Hove, Sussex.
Thurstone, L. L. (1927). Psychophysical analysis. *American Journal of Psychology*, **38**, 368–89.
Truxal, J. Q. (1972). *Introductory System Engineering*. McGraw-Hill, New York.
Weiffenbach, J. M. (1989). Assessment of chemosensory functioning in the aging: subjective and objective procedures. *Annals of the New York Academy of Sciences*, **561**, 56–64.

9

Three Generations of Sensory Evaluation

ROBERT L. MCBRIDE
Sensometrics Pty Ltd, 357 Military Road, Mosman, NSW 2088, Australia

INTRODUCTION

The following is a personal overview of the evolution of sensory evaluation. The 'three generations' to be cited are — perhaps presumptuously — categories of my own making. The main aim of the review is to point up some deficiencies in traditional methodology and suggest an alternative approach for future research.

Over the last 50 years, there have been spasmodic attempts to cajole psychophysics out of the academic laboratory and into the real world. The eminent psychologist L. L. Thurstone was perhaps the pioneer here. He clearly saw the enormous potential for psychophysics and psychological measurement to solve practical problems, and became increasingly involved in such issues throughout his career (see Thurstone, 1959). His research included perception of ethnicity, attitude scaling, prediction of choice, voting behaviour — even the sensory evaluation of food (Thurstone, 1950). In talking of the transition from Fechner to Stevens, psychophysicists are inclined to overlook Thurstone's contribution whereas it was, in reality, pioneering research of the highest order.

But despite Thurstone's auspicious beginning, adaptations of psychophysics to practical problems have been disappointingly sparse. Approaching the 21st century we find that nominal data (i.e. polling, counting heads) still suffice for most social measurement, and that many of Thurstone's techniques have yet to be implemented. Indeed, with the exception of a minority (some of whom are contributors to this volume), psychophysicists have studiously ignored issues in the real world.

This chapter is also about technology transfer. Specifically, it shows how psychophysics can bridge the gap between the technical (R & D) and marketing operations in the food industry (McBride, 1989). It is easy to see why, conceptually at least, psychophysics should work here: its very *raison d'être* is to link the *psychological* (consumer perception, marketing responsibility) with the *physical* (product characteristics, R & D responsibility). When the world is conceptualised in a psychophysical framework, the gap between R & D and marketing melts away.

To begin, it is useful to contrast these two types of research in the food industry (this will also apply to other consumer product industries). The first, technical R & D, is run by chemists, food technologists, microbiologists — usually people with training in the physical or biological sciences. These people think and talk in *physical* units such as °Brix, total solids, pH and plate counts. With focus on the product, they will have an intimate knowledge of formulation, shelf-life, flavour stability, and so on, but they will have no clear idea of how these physical parameters relate to consumer acceptance.

The second type is market research, an operation more likely to be run by people with backgrounds in the behavioural sciences. These people talk in terms of subjective responses, preferences, perceptions; they are interested in consumer reaction to a product rather than the product itself. In fact they often have no knowledge at all of what constitutes a product — of how it is made or what is in it.

The schism between these two types of research can obviously present difficulties in the organisation and coordination of a commercial operation. Not the least of these is factionalism, a result of the human propensity to align with other like-minded individuals. Thus, the technical person's stereotype of marketing may be of a glib, fickle enterprise, with style but little substance. Marketing, on the other hand, may see R & D as technically competent and necessary, but at the same time insular, lacking spontaneity, and unworldly.

As a science, sensory evaluation is about 50 years old (Amerine *et al.*, 1965) and has the potential to link the two, as noted above. The question is, has it done so in practice? My view is that for most of its 50 years, sensory evaluation has not acted as a link at all: either it has been used independently by R & D and by marketing for their separate purposes, or, alternatively, it has functioned as an adjunct to R & D alone, largely failing to advance the link with consumer acceptance. We might call these alternatives 'first generation' and 'second generation', respectively.

FIRST GENERATION

The trademark of first generation sensory evaluation is the utter lack of any coordination between R & D and market research (Fig. 1). A hypothetical example will demonstrate.

Suppose, in the interests of cost reduction, a company wishes to switch suppliers of an existing ingredient. The major concern for R & D is that the new variant may cause a change in flavour. So, R & D conducts sensory evaluation: the two samples are submitted to a taste panel using some type of difference test (e.g. triangle test, duo–trio test) which reveals whether or not the new ingredient renders a perceptible change in the sensory properties of the product. This is a common and legitimate application of sensory evaluation; note, however, that it is for the self-contained purposes of R & D. Were a difference obtained, R & D would probably reject ingredient X, even though, possibly, consumers might prefer the new formulation. R & D would not know this, because they do not conduct consumer trials.

Turning now to market research, suppose a marketing department wishes to know which product variant consumers prefer, A or B. This time, sensory evaluation takes the form of a paired-comparison taste test with consumers. A preference result is obtained and fed back to the marketing department. This exercise is also self-contained. The result will serve its purpose but the market researcher, with little or no product knowledge, may have no clear idea *why* the preference came out the way it did. It might possibly be pinned to a simple difference in product formulation, but the market researcher would not know how to do this.

These uncoordinated, independent forays into sensory evaluation are still common in the food industry: R & D does the product development work and laboratory sensory testing, then the variants are passed on to market research for consumer testing. There is no attempt

Technical R & D	Consumer Research
sensory evaluation	sensory evaluation
e.g. difference tests	e.g. preference tests

FIG. 1. Schematic representation of first generation sensory evaluation.

to link the consumer response to the laboratory sensory panel, or even more pertinent, to the product formulations.

SECOND GENERATION

The second generation approach recognises the desirability of a link. And the way to accomplish this is seen to be via the techniques of *descriptive analysis*, a major stream of sensory activity which began to appear in the 1950s (e.g. see Amerine *et al.*, 1965; Lawless, this volume; O'Mahony, this volume; Stone & Sidel, 1985).

Williams *et al.* (1984, p. 235) succinctly outline the logic of this approach: 'analytical information (i.e. the physical parameters from R & D) must be related to and interpretable in terms of consumer likes and dislikes. Unfortunately, because of the complexities of food acceptance, making direct link-ups between consumer and chemical data is fraught with difficulties. The only sensible way to proceed in unravelling these interrelationships, therefore, is to use objective sensory information (i.e. descriptive analysis) as the link between the two'. — This schema is shown in Fig. 2.

Although this seems sensible on paper, I believe that descriptive analysis has not functioned well as a link in practice. There are several reasons.

First, there is the insularity of the descriptive analysis operation. Sensory technologists have usually trained in the applied physical sciences, such as food technology. A recent survey conducted in the United States (Anon., 1988), revealed that less than 5% of sensory professionals had received education in psychology or marketing. As a consequence, their focus and interest is more R & D related; they talk the same language as the product developer but they are not necessarily on the same wavelength as the marketing executive. They are apt to become immersed in sensory profiling as an end in itself, seemingly oblivious to the fact that a better understanding of consumer acceptance should be a primary goal.

A second problem is one of which sensory practitioners are well

Technical R & D	Sensory evaluation	Consumer research
product development	descriptive analysis	consumer preference

FIG. 2. Schematic representation of second generation sensory evaluation.

aware — semantic ambiguity in profiling. What is 'fruity' to me not be 'fruity' to you. Sensory psychologists have been slow to grapple with these problems in concept formation, although research is now appearing (Ishii & O'Mahony, 1987; Lawless, this volume; O'Mahony, this volume). Statistical manipulations, such as procrustes analysis, have been developed in an attempt to overcome these hurdles, but these, too, suffer shortcomings (Huitson, 1989).

From an epistemological standpoint, there is a more serious shortcoming of descriptive analysis — it is descriptive, not explanatory. While this issue might be overlooked by a pragmatic food technologist, it cannot be evaded by the scientist. Profiling cannot itself tell us *why* consumers prefer A over B, nor can any amount of statistical manipulation. Because some of the multivariate analyses now available are, statistically speaking, very sophisticated, it is tempting to imbue them with explanatory power. But they do not have explanatory power. I must say that complex statistical treatments have never advanced my understanding of sensory processes; on the contrary, they sometimes seem to be 'science upside down'. The investigator uses the analysis to resurrect order from a data matrix, the disorder of which was caused by lack of experimental design in the first place. Arbitrary meaning is attached to the 'dimensions' *after* the data analysis, when these dimensions might have been set up unequivocally in the initial design!

To be more specific, descriptive analysis is phenomenological: it is concerned only with the subjective world of sensations and fails to relate these to physical parameters on the one hand, and to hedonic tone on the other (i.e. to the left- and right-hand sides of Fig. 2). In regard to the link with the physical, the main problem is that investigators have 'jumped in at the deep end'. Instead of beginning with simple stimulus systems and building a database on the sensory interactions therein, they have proceeded directly to complex stimuli in the real world (e.g. a selection of wines, or coffees, or orange juices). The commercial appeal of this tack is understandable, but the stimulus complexity makes establishment of a psychophysical link all but impossible.

Moreover, the other link, between sensory characteristics and consumer preference, has received even *less* attention. Unlike difference testing, for which there is a well-established theoretical framework (signal detection theory), acceptance testing has almost no theoretical base at all. This is a curious blindspot given that hedonics have primacy in eating, and that the food business is the largest in the world.

With these drawbacks, we might question the central premise of the second generation approach: is it really necessary to invoke descriptive analysis as an intermediary? Might this not be an unnecessary complication? After all, the accept/reject response is in no way language dependent; it is observed in the newborn (Desor *et al.*, 1973), and infants develop decided preferences long before they can talk. Even in articulate adults the hedonic response is still primary, the use of descriptive language being a secondary operation.

As an aside here, it should be noted that the greater the degree of liking, the sparser the descriptive terms available. There is no shortage of descriptors to label defects in a product, but when a food is acceptable — especially when it is highly acceptable — consumers find themselves limited to variants of hedonic terminology, such as 'superb', 'well-balanced', or 'just right'. It is, literally, too good for words. In this event, why not bypass the descriptive phase altogether?

THIRD GENERATION

In this approach, the consumer response is directly linked to its stimulus (Fig. 3). There is nothing novel about this, of course. It is simply traditional psychophysics: vary the stimulus (cause) in a systematic, controlled manner, and measure the response (effect). The response in this case may be either intensive or hedonic — in either case, however, there is no intervening descriptive analysis.

A shortcoming of traditional psychophysics in this role has been that it is univariate. It deals with one parameter at a time, whereas the real world, with real products, is multivariate. People do not eat or drink univariate stimuli such as sugar in water. Even lemonade, a relatively simple product, consists of three main components (lemon flavouring, sugar and acid). So, we need a way to accommodate stimulus complexity and find out how all aspects of the stimulus information are integrated into the acceptance response.

Integration psychophysics is one way to accomplish this end (McBride & Anderson, this volume). In this approach, power comes from the

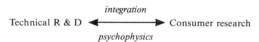

FIG. 3. Schematic representation of third generation sensory evaluation.

experimental design. Trained taste panels are not needed; the research can be done with novice consumers in many cases, with 'smart' consumers in others. An important advantage of integration psychophysics is that it does not rely on verbal reports of what consumers *think* are the important components in a food or drink (McBride, 1989). As we shall see in the following examples, this can be misleading.

What makes an orange drink 'orangey'? An orange drink is a simple formulation, the major components being colouring, orange flavouring (supplies the orange aroma), sugar (sweetness) and acid (sourness). Conventional wisdom would have it that the orange flavouring is the most important contributor to oranginess. But is this really so?

The key is to use a factorial design in which the relevant components are varied independently. In this simple example (McBride, unpublished data) three levels of orange flavouring were varied with three levels of sugar in a constant citric acid background. Colour was not used. All samples were presented to a consumer taste panel who knew nothing of the manipulation; they simply tasted the drinks and rated them for intensity of orange flavour.

The results are surprising, even to a food scientist. The bottom curve in Fig. 4 gives the perceived intensity of orange flavouring (at three levels) when mixed with a constant base of citric acid but with no sugar. This curve shows that perceived orange flavour does indeed increase with the concentration of orange flavouring in the drink.

However, the perceived orange flavour increases even more markedly with the addition of the first level of sugar (middle curve) and further still with the second level of sugar (top curve). The plot shows, in fact, that sugar had a larger effect on perceived orange flavour than the orange flavouring itself! The high level of sugar alone, with no orange flavour (leftmost data point on top curve) made the drink more 'orangey' than a high level of orange flavouring without sugar (rightmost data point on bottom curve). The most efficient way to increase oranginess is therefore not to add more orange flavouring, but to add more sugar.

This somewhat counter-intuitive result is commercially useful as well as scientifically interesting. Note that it was obtained from a straightforward consumer response, not a sophisticated flavour profiling; the analysis is simple (in fact mere inspection of the pattern in Fig. 4 will suffice); and, most importantly, because the stimulus was varied systematically and the resulting response measured, the relationship is genuine cause-and-effect, not merely correlational.

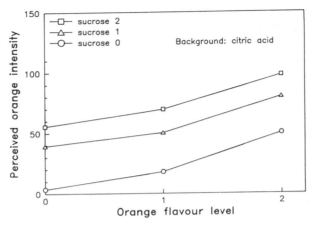

FIG. 4. Perceived oranginess of mixtures of sugar and orange flavouring in a base of citric acid. The three levels of orange flavouring are given on the x-axis and each curve corresponds to a different level of sugar.

More fundamentally, the human inability to sense the relative importance of components suggests that hedonic processing is preconscious. This would also explain the dearth of positive descriptive terms (aside from hedonic) in language: obviously it is impossible to develop descriptive terms for events which are not conscious.

So, sweetness is the principal component of orange flavour. Having identified this principal component, it remains to determine exactly how much sugar should be used. As in most food products, there will be an optimum level of each ingredient, in this case sugar, below and above which the acceptability will decline. This optimal level can be determined in a number of ways (e.g. McBride, 1982; McBride & Booth, 1986) which will not be elaborated here, except to point out that, in addition to estimating the optimum level of an ingredient, appropriate analysis will provide the deviation from that level that is just tolerated by consumers (Booth, this volume). In this way, simple acceptance data from a sample of consumers can be fed back to manufacturing and incorporated in the quality assurance program.

The case of the tasteless tomatoes. Over recent years, Australian tomatoes have consistently been the target of consumer complaints. Comments such as 'tasteless' and 'flavourless' continually appear in the popular press. For the sensory researcher the task is to discover exactly what component of tomatoes is amiss; however, given the previous example, we cannot expect the consumers themselves to know what it is. Is it taste or aroma? And, if taste, which taste?

Unlike the orange drink, tomatoes are a natural food and their sensory components are not readily manipulable in terms of experimental design. What can be done, however, is to take several varieties of tomatoes which differ naturally in sensory composition, present these to consumers in a controlled manner, and measure the hedonic response.

To discern the relative importance of aroma versus taste, we ran consumer evaluation of each variety under three separate conditions: (1) aroma only (sniffing tomato samples without tasting); (2) taste only (eating tomatoes with the nose pinched, thereby preventing the aroma volatiles reaching the receptors in the nasal epithelium); and (3) full flavour (normal eating). The responses in these three conditions allow the relative importance of the sensory attributes to be inferred, albeit not as cleanly as with the orange drink.

Aroma proved not to be an important factor: there was virtually no difference in aroma between varieties of tomatoes, and besides, preventing perception of the volatiles did little to impair acceptability of the tomatoes. Texture, likewise, did not appear to play a role in acceptability inasmuch as there was little difference between varieties.

We deduce that taste is the important contributor, but which taste, sweetness or acidity? In fact, sweetness proved to be of predominant importance. The higher the sugar, the more the tomatoes were liked. Significantly, consumers were not conscious of the role of sweetness: a tomato higher in sugar was described as a 'good tasting tomato', not a 'sweeter' tomato. Conversely, unacceptable tomatoes were never identified as 'not sweet enough'. This is consistent with sensory integration as a preconscious operation.

A conceptually similar case study is reported by Fishken (1988). Qualitative research on a brand of pizza indicated a desire for more cheese and more meat. However, when actual taste tests were run, with formulations varied systematically according to factorial designs, what consumers *thought* they wanted was not what they wanted in reality. The taste tests revealed a liking for more cheese and *less* meat.

There is a salutary lesson here for those who rely heavily on qualitative (focus group) data. Qualitative techniques can be useful for testing the viability of new concepts, but they are inadequate whenever sensory change is a possibility.

Two contrasting examples will demonstrate. First, suppose we wish to measure the consumer reaction to 'low cholesterol' foods. Consumers cannot see, feel, smell or taste cholesterol, so in this case a concept test is adequate. The only way to tell if a product is low in cholesterol is to read it on the label.

Now suppose we test consumer reaction to 'low salt' foods. This time the predictive validity of concept testing will be shaky, because a reduction in salt level will change the sensory characteristics and hence the acceptability. A favourable concept reaction can be overriden by a negative product (hedonic) reaction.

A STRATEGY FOR SENSORY EVALUATION

I believe the future for sensory evaluation is bright provided that it continues to expand its theoretical base. In any field of endeavour, descriptive research eventually becomes banal, and this is a risk for sensory evaluation.

It is ironic that although hedonics drive food acceptance, theory construction here is still in its infancy. Indeed, even data collection is in its infancy; and we cannot expect to develop a theoretical structure for hedonic psychophysics without an empirical database of sensory interactions and integration.

Hedonic research does not fit neatly into established research categories, which has been another hindrance to progress. It sits in that grey area between perception and motivation, not well accommodated by tertiary education. Experimental psychology tends to be strong in perceptual (but not hedonic) theory, whereas food science, although more in touch with the practical importance of hedonics, is not strong on theory construction. There is an opportunity for consumer psychology to fill this gap.

REFERENCES

Amerine, M. A., Pangborn, R. M. & Roessler, E. B. (1965). *Principles of Sensory Evaluation of Food*. Academic Press, New York.

Anon. (1988). SED survey results. IET Sensory Forum, No. 41, 1, 3.
Desor, J. A., Maller, O. & Turner, R. (1973). Taste in acceptance of sugars by human infants. *J. Comp. Physiol. Psychol.*, **84**, 496–501.
Fishken, D. (1988) Marketing and cost factors in product optimization. *Food Technol.*, **42**(11), 138–40.
Huitson, A. (1989). Problems with Procrustes analysis. *J. Appl. Statist.*, **16**, 39–45.
Ishii, R. & O'Mahony, M. (1987). Taste sorting and naming: Can taste concepts be misrepresented by traditional psychophysical labelling systems. *Chem. Senses*, **12**, 37–51.
McBride, R. L. (1982). Range bias in sensory evaluation. *J. Food Technol.*, **21**, 405–10.
McBride, R. L. (1989). Psychophysics — A link between R & D and marketing. *Food Technol. NZ.*, **24**(7), 25–7, 30.
McBride, R. L. & Booth, D. A. (1986). Using classical psychophysics to determine ideal flavour intensity. *J. Food Technol.*, **21**, 775–80.
Stone, H. & Sidel, J. L. (1985). *Sensory Evaluation Practices*. Academic Press, London.
Thurstone, L. L. (1950). Methods of food-tasting experiments. *Proc. 2nd Conf. Res. Am. Meat Inst.*, pp. 85–91.
Thurstone, L. L. (1959). *The Measurement of Values*. University of Chicago Press, Chicago.
Williams, A. A., Rogers, C. & Noble, A. C. (1984). Characterisation of flavour in alcoholic beverages. In *Flavour Research of Alcoholic Beverages*, ed. L. Nykanen & P. Lehtonen. Foundation for Biotechnical and Industrial Fermentation Research, Helsinki.

Index

A-not A test, 120, 125
Acceptance testing, lack of theoretical base, 199
Adaptation
  cross, 49–50
  defined, 46
  eating, and, 47–9
  locus, 49
Aging
  effect on sensory function, chemosensory preference, 26–32
    nutrition and chemosensory perception, 32–5
    olfaction studies, 24–6
    taste, 20–3
Amino acid preference, and nutritional status, 34
Analysis vs integration, 42–4
Appearance–flavor studies, 105–6
Apple juice mixtures, studies, 111
Apples, hormonal sprays, 112–13
Attitudes
  beliefs, and
    conceptualisation, 143
    food choice, 141–6
  changes, 157
  expectancy value model, 148
  Fishbein and Ajzen model, 146–57
    modification, 155–8

Attitudes—contd.
  motivation to comply, 148
  normative beliefs, 148
Audition
  analysis and synthesis, 43–4
  pitch discrimination, 45
Auditory pathway, tonotopic organization, 43

Behavioural intention, equation, 147
Beliefs. See Attitudes, and beliefs
'Best-before' dating, 187
Bias-permitting tests, 185
Bitterness, age-related losses, 21–3
Blood urea nitrogen, and nutritional status, 33
Bread, low-salt, 149
Butanol, olfactory threshold, 24
Butyric acid, biased response in children, 9

Caffeine
  mixtures, age-associated threshold loss, 21
  recognition, 62–3
Casein hydrolysate, effect of aging and biochemical status on preference, 33–5
Chemosensory perception
  effect of age, 26–32
  nutrition, and, 32–5

Children
  black/caucasian, salt preferences, 12–14
  taste perception
    dietary experience, 9
    preference technique for preschoolers, 9–14
    preferential ingestion of salted food, 8
    review of the literature, 7–9
Cholesterol in food, 204
Citric acid
  age-associated threshold loss, 21
  pleasantness ratings, 28–30
  salt, and, 51
  studies in infants, 5
  sugar, and, 51, 96–7, 107–9, 110
Coffee
  taste–odor integration, 106
  see also Caffeine
Color
  Munsell Book of Color, 130
  perception, 129–30
  relationship with taste, 41–2
Concept formation, descriptive analysis, 129–30
Consumer panels
  measurement of objective sensory preferences, 181–5
  numbers, 182
  preference in sensory testing, 185–7
  trained vs novice, 201
Consumer science
  basis, 167–70
  causal structure of preference, 173–5
  instrumental and process factors in preference, 170–3
Context effects, 79
Controlled Preference Test, 187
CR_{max}, in tasting, 59

Descriptive analysis
  concept formation, 129–30
  in the food industry, 130–2
  measurement of sensory

Descriptive analysis—contd.
  concepts, 132–4
Design information, personalised, 163–7
Diabetes, and sweetness preference, 26
Difference tests. See Sensory difference tests
Dominant component rule, 101, 108
Dual standard test, 119
Duo-trio test, 119–20, 125

Eating
  adaptation, and, 47–9
  see also Taste
Expectancy-value model
  attitudes, 147
  equation, 147

Facial Action Coding System, 4
Facial expressions, neonates, 3–4
Fallis-Lasagna-Tetreault test, sensory difference, 120
Fast-food restaurants, and behavioural intentions, 148
Fats
  consumption, attitude, 150–1
  see also Milks
Fishbein and Ajzen model, attitudes, 146–57
Flavor Profile Method, 131
Flavor–appearance studies, 105–6
Food choice
  attitudes, and beliefs, 141–6
  Fishbein and Ajzen model, 148–57
  hedonic reactions, 27, 157
  influencing factors, 142–3
    relative importance of factors, 145
  and nutritional practice, 144
  and specific sensory attributes, 144

Food industry, and descriptive analysis, 130–2
Food industry, concept alignment
   Quantitative Descriptive Analysis, 131
   Spectrum Method, 131
'Forced-choice' test, 119, 124–5
   staircase procedure, 24

Gestalt psychology, 43
Glucose, neonates, adaptation to glucose solutions, 2
Gusto-facial responses, neonates, 3–4
Gymnema sylvestre extract, 54–5

Habituation, and taste, mixture suppression, 55–6
Harris–Kalmus test, 120, 128
Hedonic reactions
   food choice, and, 156–7
   integration, 111–13
   neonates, 1–2
   peak preferred concentration, and, 27
   quantitative pleasantness judgment, 27
   terminology, 200
   tone, 109–11
      intensity information, and, 27

Ideal-relative descriptive ratings, 189–91
Ideal-relative profiling, 188–9
Information integration theory
   conceptual advantages, 100–1
   dominant component rule, 101, 108
   empirical studies, 105–13
   integration function (I), 94
   logic,
      conceptual basis, 94–7
      experimental procedure, 99–100
      parallelism theorem, 97–9

Information integration theory—*contd.*
   mixture analysis, 100
   nonmetric stimuli, 101
   prescribed rules, 104–5
   subtractive threshold rule, 97, 101
   technical advantage, 102–5
   vs analysis, 42–4
Information processing in taste
   modest abilities, 60
   number of components, 61–2
   onset of stimulus components, 62–3
   summary, 63
Integration psychophysics
   a conceptual shift, 113
   logic, 94–100
   see also Information integration theory

Judges, need for training, 131
Just Noticeable Difference (JND), defined, 22

Linear fan theorem, 101
Linear response scales, 99

Magnitude estimation
   linear response, 99
   rating method, 99–100
   taste suprathreshold intensity, 20
Magnitude matching, 20
Market-realistic test situations, 187
Meat, choice, 149
Memory, in consumer research, 83–4
Milk
   choice, 149
   low-fat, 152–5
Mixtures
   heterogeneous, 107–9
   homogeneous, 106–7
   mixture analysis, 100–1
   suppression, response patterns, 96–7
Motivation to comply, attitudes, 148
Multiple samples, sensory difference tests, 127–8

Nasal airway resistance changes, and olfaction, 24–5
Neonates
  facial expressions, 3–4
  hedonic reactions, 1–2
  sucking reflex *vs* taste stimulus, 5
  taste perception, 1–7
Normative beliefs, attitudes, 148
Null hypothesis, and panel agreement, 85
Nutrition
  chemosensory perception, and, effect of age, 32–5
  deficiencies in old age, 32–4
  food choice, and, 144
Nutritional status
  amino acid preference, and, 34
  defined, 33

Octad test, 120, 128
Odor mixtures, 109
Odor–taste studies, 106
Olfaction
  early development of sensitivity, 3
  effect of age, 24–6, 32
  nasal airway resistance changes, and, 24–5
  olfactory integration, 109
Orange drinks
  formulation, 201–2
  hedonic integration, 111

Paired comparison test, 119
Panel agreement and null hypothesis, 85
Perception. *See* Sensory perception
Personalised design information, 163–7
Pizza, taste preferences, 202
Pleasantness ratings, ANOVA, 28
Preference, causal structure, 173–5
Preference and sensory testing
  Controlled Preference Test, 187
  hedonic testing, 185–6
  improvement of existing techniques, 187–91

Preference and sensory testing—*contd.*
  difference tests, 189–91
  optimisation, 188
  market-realistic test situations, 187
  measurement,
    analysis using multiple regression, 184–6
    bias-permitting tests, 185
    consumer panels, numbers, 182
    product sample selection, 182–4
    tested state and situation, 181–2
  peaked preference, 173–4
  preference scores, measurement in consumer science, 170–3
  representative consumer panel, 186
  selection but not training, 186
  strength of an objective influence, 175–81
  tolerance triangle,
    aggregation of behaviour, 181
    analysis of data, 174–5
    intercepts, 176
    preference slope, 177–8
    residual variance, 178–80
    tolerance line, 175–6
  see also Sensory evaluation
Process factors, 170–3
Product design
  basis of consumer science, 167–70
  customer's description, 164
  instrumental and process factors, 170–3
  open-ended questioning, 164
  segmental positioning, 164, 165
  social statistics and individual cognitive science, 165–7
Product sample selection, 182–4
Preference and sensory testing. *See also* Sensory evaluation

Quinine
  salt, and, 58

Quinine—*contd.*
  sucrose, and, 51, 57–8
  sugar, and, 111–12
Quinine sulfate, neonates, 4–6

R index, 123, 134
Reaction time experiments, 83
Receiver Operating Characteristic Curve, 123
Receptor sites, occupancy, and adaptation, 49
Response function (M), 94
Response patterns
  mixture suppression, 96–7
  taste–odor integration, 95–6

Salt
  adaptation curves to saltiness, 47, 55
  citric acid, and, 51
  discrimination thresholds and aging, 20–1
  enhancing effect on sugar, 50–3
  low salt foods, concept *vs* hedonic reaction, 204
  meat, and, 150
  neonates,
    adaptation to saline solutions, 2
    gusto-facial responses, 4–5
  pleasantness ratings, 27–32
  preferred levels,
    black/caucasian children, 12–14
    children, 10–11
  quinine, and, 58
  sugar mixtures, and, 28, 50–63
  table salt use, 151–2
  water *vs* beverage base, 29
Sensory concepts, measurement, 132–4
Sensory difference tests
  A–not A test, 120, 125
  criterion variation,
    discussion, 120–1

Sensory difference tests—*contd.*
  forced choice procedures, 119, 124–5
  signal detection procedures, 122–4
  single stimulus judgements, 125–9
  defined, 117
  difference thresholds and scaling, sensory evaluation, 71–5
  difficulty of administration, 189–91
  dual standard test, 119
  duo-trio test, 119–20, 125
  Fallis–Lasagna–Tetreault test, 120
  'forced-choice' test, 119, 124–5
  Harris–Kalmus test, 120, 128
  Octad test, 120, 128
  paired comparison test, 119
  tetrade test, 120, 125–56
  triangle test, 120, 125
Sensory evaluation
  definition, 79
  descriptive analysis, 198–200
  difference testing, difference thresholds and scaling differences, 71–5
  effect of age, 26–32
  fundamental questions, 69–71
  identification of important sensory attributes, 75–9
  integration psychophysics, 200–4
  learned concepts, as, 78
  measurement of concepts, 132–4
  methods research, 84–9
  overview, 195–7
  perceptual issues, memory and context, 79–84
  R & D and market research, 197–8
  ratings, 80
  strategy, 204
Sensory function, effect of age, 20–3
Sensory evaluation. *See also* Preference and sensory testing
Smell
  part played in eating, 48

Smell—*contd.*
  taste, and, factorial plot, 96
Subtractive threshold rule, 97, 101
Sucking reflex
  burst length, 6
  vs taste stimulus, 5
Sugar
  citric acid, and, 51, 96–7, 107–9, 107–10, 110
  diabetes and sweetness preference, 26
  discrimination thresholds and aging, 20–2
  mixtures of homogeneous substances, 106–7
  mixtures with salt, 28, 50–63
  pleasantness ratings, 27–32
  quinine, and, 51, 57–8, 111–12
  suprathreshold intensity in aging, 20–2
  sweetness perception, age-related losses, 21–3
  orange flavour, and, 201–2

Taste papillae, loss, 22
Taste perception
  absolute gustatory sensitivity, 2
  color, and, 41–2
  complex stimuli, 41–42
  effect of age, 20–3, 32
  Flavor Profile Method, 131
  flavor–appearance studies, 105–6
  heterogeneous mixtures, 107–9
  homogeneous mixtures, 106–7
  information processing, 60–3
  like–dislike scores, 42
  mixture suppression,
    bilateral stimulation of tongue, 56–60
    central/peripheral origin, 53–5
    conditions affecting, 51
    habituation, and, 55–6
    mixture enhancement, 52–3
    side tastes, 52–3
  neonates, 1–7
  odors, and, 95–6, 106

Taste perception—*contd.*
  older infants and young children, preference technique for use with preschoolers, 9–14
  review of the literature, 7–9
  parallelism theorem, 97–9
  perceptual integration, and analysis,
    attention, 44–5
    food perception, 43–4
    Gestalt psychology, 43
    sensory analysis, 45
    simple and complex stimuli, 45–6
    studying *vs* eating and drinking, 46
  smell, and, factorial plot, 96
Taste–odor integration
  response patterns, 95–6
  studies, 106
Tetrade test, 120, 125–56
Thurstonian scaling, 87–8
Tolerance triangle, preference and sensory testing, 174–5
Tomatoes, taste preferences, 202
Tongue
  aging effects, 22
  bilateral stimulations, 56–60
  innervation, 57
  structure and taste function, 47–8
Triangle test, 120, 125
Trigeminal sensitivity, 19

Urea, bitterness testing, 5–6

Validity, evaluation, 86–7
Valuation function (V), 94, 101

Weber Ratio
  defined, 22, 48
  use in discrimination threshold testing, 22–3
Weight loss, behaviour, 150

WITHDRAWN

Books are to be returned on or before the last date below.

THIS ITEM MUST BE RETURNED TO THE SITE FROM WHICH IT WAS BORROWED

1-DAY LOAN

SHORT LOAN NO RENEWAL

LIBREX